Deep Disagre ___t in U.S. Agriculture

Deep Disagreement in U.S. Agriculture

Making Sense of Policy Conflict

Christopher Hamlin
and Philip T. Shepard

Routledge
Taylor & Francis Group

LONDON AND NEW YORK

First published 1993 by Westview Press

Published 2018 by Routledge
52 Vanderbilt Avenue, New York, NY 10017
2 Park Square, Milton Park, Abingdon, Oxon OX14 4RN

Routledge is an imprint of the Taylor & Francis Group, an informa business

Library of Congress Cataloging-in-Publication Data
Hamlin, Christopher, 1951–
    Deep disagreement in U.S. agriculture: making sense of policy
conflict / by Christopher Hamlin and Philip T. Shepard.
    P.   cm.
    Includes bibliographical references and index.
    ISBN 0-8133-8703-5
    1. Agriculture and state—United States.   I. Shepard, Philip T.
II. Title.
HD1761.H358   1993
338.1'873—dc20

ISBN 13: 978-0-367-01171-0 (hbk)
ISBN 13: 978-0-367-16158-3 (pbk)

Contents

Figures

Abbreviations
of Common Citations

Preface

This is a book about disagreement, particularly the kinds of disagreements that occur today in making policies for science and technology. We use the area of agriculture—agricultural research and agricultural policy—to exemplify such disagreements because it so richly reflects their range and depth.

Not only is such disagreement commonplace, it is much studied: we know much about alternative philosophies of medicine, that people have a great range of views on environmental and energy policies, and so forth. Indeed, differences of opinion are often seen as a normal and even healthy condition of society. Not only is it well known that a great deal of difference exists; we often think that we know why. People have different views because they occupy different stations in society or have different cultures and backgrounds; they have different interests to defend, different sensibilities and perceptions, or different values and beliefs, some of which may go with their social territory. Such perspectives have prompted a great explosion of writing in the sociology of science and technology in recent decades.

Yet from another point of view it is really quite puzzling that people disagree so much, particularly people within a given culture. In regards to American culture this puzzlement is especially acute since the culture seems to place such great faith in rational means of decision making and seems to draw so heavily on Western traditions of analysis and problem solving that have been cultivated for centuries. After all, the methods—whether of the ethicist, jurist, debater, or scientist—are intended to resolve disagreement. How can it be that the more we rely on these methods, the more and deeper our disagreements seem to grow? Are the causes of disagreement so fundamental that reason cannot touch them? that appeals to reason are but deceitful ploys, aimed to legitimate and assert one's own interests above others? How is it that we manage to articulate so much disagreement even while using tools invented to guide us to agreement? One can say that people with different "values" will use the tools differently, but that only takes us in a circle, for we rarely look closely enough to distinguish the values from the disagreements as a whole.

Indeed, our American faith in reason seems to be matched and perhaps canceled out by a sort of complacency—a willingness to accept
disagreements as inevitable and incorrigible. Along with the tradition of
American individualism has come a sort of disengaged tolerance (or
sophomoric relativism) that belies our need to live together and that may
not be altogether a good thing. In many policy areas, including
agriculture, it has impeded the making of coherent or long-term policies
and has led to a politics in which manipulating the agenda is often
treated as more important than reaching an understanding. The
assumption is that one can reasonably believe whatever one wants to; but
the price of such freedom is a heavy one—for when everything is
reasonable nothing reasons, nothing serves us to make well the collective
or shared choices that we must take or to bridge the differences that
divide us and tear apart the social bonds that make it possible to live
well together.

The answer, we believe, is not so much to constrict what can be
considered legitimate disagreement as to build stronger means for
understanding how people manage to disagree, especially when their
disagreements are far from ephemeral—when they are deeply felt,
urgently expressed, and acted out in earnest. We need better ways of
recognizing and representing the kinds of frameworks of belief and
action which people bring with them into the public arena—the languages they use; the sensibilities, assumptions, and modes of argument
they draw upon; the kind of authority they claim for their statements.
Ideally, as citizens concerned with agriculture, such approaches would
allow us to consider more fully and fairly the widening range of voices
in public debate, while at the same time avoiding the facile acceptance of
all statements and forms of discourse as equally reasonable or equally
legitimate whether we understand them or not. In short, our hope is to
expand the capacity for critical discussion on matters of agriculture, not
by ruling out some participants or denouncing some arguments, but by
finding ways to work with multiple frameworks so that people operating
within different ones can understand each other better.

This project was made possible by a grant from the National Science
Foundation Program on Ethics and Values in Science and Technology
(RII-8409919). Any opinions, findings, and conclusions or recommendations expressed here are those of the authors and do not necessarily
reflect the views of the Foundation. We are grateful for this support and
for additional support from Michigan State University (College of Natural
Science and Lyman Briggs School) and The University of Notre Dame,
which allowed us to bring the project to completion.

The project has gone on for many years. A great many people have
graciously considered and criticized various parts of it in various stages

of its evolution. However fully or sparsely they may find their concerns reflected in the final manuscript, the points they made were invariably well taken, and led us repeatedly to rethink and refine various parts of the work. Among these people are: Gerald Berk, Richard J. Bernstein, Frederick H. Buttel, Kenneth A. Dahlberg, David B. Danbom, Stanislaus J. Dundon, Russel W. Erickson, Craig K. Harris, Richard R. Harwood, Richard P. Haynes, Glenn L. Johnson, Les Levidow, John Lyon, Robert McKinley, Everett Mendelsohn, Denton E. Morrison, Alven Neiman, Basil O'Leary, Dorinda Outram, Norman Pollack, Vernon W. Ruttan, Joel Schor, Harold Schwarzweller, Charles Walters jr, Ernie Yanarella, and I. Garth Youngberg. We wish to thank all of these people, any others we may have inadvertently omitted, and especially those who committed considerable amounts of their time to careful and thorough critiques of substantial drafts of the book. Their thoughtful criticism has made this a much better work than it would otherwise have been. We also thank Sharon Powers for access to sources otherwise unavailable.

Christopher Hamlin   Philip T. Shepard

PART ONE

Assaying the Problem

1

Introduction

Many people today, including the authors, are worried about the state of agriculture in the U.S., wondering where it is going and whether we will like it when it gets there. We are told that agriculture is progressing or that it is deteriorating, that it is in the midst of a crisis or simply undergoing a normal adjustment, that the future will be satisfactory beyond imagination or tragically and unspeakably wretched. Which claims are true, we wonder. Can a desirable future be achieved, and, if so, what forms of agricultural policy, and especially of agricultural science and technology, will bring that future?

Discussion, debate, and political interaction on such agriculture-related issues has often been highly polarized. Consider some common observations about current controversies in agriculture. The participants, it is often noted, talk past one another. Their use of key terms, their assessments of what the problems are, and their versions of the facts vary widely. Their guiding visions of what agriculture could and should be, and their standards for judging what is done in agriculture are greatly at odds. Their programs and goals for research diverge, sometimes radically. Rarely do they address the same questions, and when they do they come at them from such different directions that one hardly knows how to compare the responses, much less determine who is right and who wrong. Opponents appear not to understand each other very well: "Where are they coming from? What do they mean? How can they possibly believe that?"—the questions are raised, but no answer is expected. The participants, in short, are operating in different frameworks (Coleman 1982: 581; Aiken 1982: 29-30). Under such conditions even limited compromises are difficult to achieve, discussions about the long term goals of U.S. agriculture seem to go nowhere (Browne 1988: 216ff).

When people talk past one another, and when policies do not take account of contesting views, political polarization often results. Opponents are driven further apart; like-minded individuals close ranks and become increasingly defensive. Failure to communicate fuels conspiracy theories, suspicion and distrust, closing off new opportunities for dialogue. As opponents separate "us" from "them," the opposition is demeaned and "we" turn inward to bolster "our" ways and ignore the problems shared with outsiders. Controversy is exacerbated, disagreements deepen, and political resources are used up with little benefit (Hileman 1990a,b; Killingsworth and Palmer 1992; Socolow 1976).

In this book we attempt to open a path to more fruitful communication among participants in agricultural controversy. We do so by mapping the outlooks of participants so that they and we can find our way across a complex terrain of assumptions, arguments, and assessments, which may include strange violations of what any one of us will take for granted in our own settled way of thinking.

Current Controversies in U.S. Agriculture

Failed communication, increasing polarization, and the formation of hostile camps—all are marks of recent controversy about U.S. agriculture. To be sure, agricultural policy has involved controversy for most of our history, first over tariffs, homesteading, and land distribution issues, then over railroad regulation and rural social services, and then over price controls (Cochrane 1979). In most of these controversies the positions were few, the concerns clear, and the possible compromises (or intractable disagreements) easily recognized. By the late 1930s a "triangle of power" had arisen (commodity groups, a congressional "farm bloc," and USDA administrators) which stabilized the policymaking process even while it reversed itself frequently on the policies themselves.

Things are different today. The structure of food production in the U.S. has changed. There are fewer farmers and their concerns and interests are more specialized. New groups have become publicly concerned with agriculture and food issues. The range of controversy has broadened and the difficulty of making policy deepened (Albrecht and Murdock 1990). New complexities have intruded on a daunting range of topics: the survival of "family farming," the importance of farming life to families and to living "in balance" with nature, the cost of food to consumers, the quality of the food supply, the rights and welfare of farm animals, the preservation of farmland and of wild margins, the ecological stability of the individual farm and the world ecosystem, the integrity of indigenous cultures, the importance of securing international markets, the ethical problems of world hunger, alternative forms of farm aid and farm finance, inheritance taxation, labor policies for farm workers, threats of food scarcity, the use of pesticides, fertilizers, and farm machinery, the organization and regulation of processing and marketing, how to improve agricultural technology, the dependency of developing countries, and the limitations of agricultural science.

The issues shift from year to year and forum to forum. Some, like the conflict over farm prices or over the demise of the family farm, are longstanding. Others, such as concerns about the regular use of antibiotics and growth hormones in animal raising, or concern about the impact of tomato harvesters, are results of technological innovation (Buttel 1986a). Still others—concerns about the health effects of pesticides, about the loss of soil fertility, and the depletion of energy resources—rise and fall periodically and have done so for centuries in some cases. With the development of technologies with radically new social and moral implications, such as genetically engineered materials, the range of issues and variety of voices is likely to increase rather than diminish (Rautenstraus 1980; Browne 1987; Lacy and Busch 1988; Gendel et al. 1990; MacDonald 1989; Baumgardt and Martin 1991).

It may be tempting to exclude many of these issues, to deny they are really agricultural issues, or to relegate them to the limbo of personal philosophy. But it is doubtful whether the agricultural establishment can politically afford to dismiss them. Groups that formerly had no standing in agricultural policy have become better organized, raised funds, and committed time and money to teaching, writing, and taking political action to make their concerns heard. Many have succeeded. Legislatures, courts, and foundations have given support to their concerns. Omnibus farm bills in 1985 and 1990 clearly reflected their power (Browne 1988; Meyers 1989; Constance et al. 1990). In any case, to understand current controversy we have to accept an enormously expanded variety of issues and groups with public voice, a diversity that extends even to the meanings of "agriculture," or more recently of "sustainable agriculture" (Bonnen 1984; Hadwiger 1980: 40-1; Hadwiger and Browne 1978; Beus and Dunlap 1990; Murdock et al. 1990; Dahlberg 1991). In short, it is no longer clear where the center of power over agricultural policymaking lies or will lie. Those who formerly had "clout" must now seek coalitions.

The Burden on Policymaking

In accepting these changes we must also recognize the overwhelming burden they have placed on the policymaking process itself. It is always difficult to win acceptance for feasible proposals when faced with conflicting views and demands. The deeper the conflicts, the harder it is.

If the conflicts are deep enough and sufficiently complex, tension in the policy process may allow only two options: a meaningless compromise or the blunt exercise of power. Yet neither option is likely to provide a durable solution.

When circumstances are favorable, policymakers within democratic traditions often find it fruitful to shape policy by open discussion and dialogue. Differences of outlook can then be tapped as resources, a wide spectrum of options can be assessed, and a workable compromise can be determined before proposals are mounted formally. Disagreement can be civil and unresolved differences manageable. But when polarization occurs these options disappear. When it gets bad enough, policymakers' options dwindle to choosing the lesser evil.

On matters so complex and so important it would seem that open discussion is worthwhile. Policymakers and interested citizens alike need to grasp the character of disagreement, locate the sources of misunderstanding, and alter the climate so that political breaks can be mended. But is it possible to bring about a political environment that supports such dialogue? That is the problem this book addresses. We believe that more fruitful dialogue across the whole range of agricultural and food policy issues can be facilitated by building bridges of comprehension that will help those concerned about agriculture understand where their sensibilities have split apart.

Responses to Controversy

How are these ends to be achieved? In particular, if civil communication is the key problem, how is one to intervene without exacerbating polarization? In this book we intervene descriptively; much of it is devoted to a description of three outlooks prominent in agricultural controversy. We do not attempt to identify which of these is best, or to discover a way of compromising among them, or even to assess their particular merits or failings. Nor do we dwell upon their location in society with an eye to exposing the motivations or interests they represent. We try only to make a sense of them.

But why do that? Why describe, why only describe? And what are these descriptions to do? To many, description without evaluation and recommendation is irresponsible. Either it smacks of ivory tower escapism, something one does to avoid the moral and political responsibility of having to act, or it is a sneaky way to smuggle in one's own views by pretending they're simply descriptions of what is.

We see description as a mode of political action, but in a somewhat unusual way. In describing opponents' deeper differences to and for each other by translating around misunderstandings, we hope to remove those causes of polarization that are grounded in sheer incomprehension (whether of another's views or even of one's own).[1]

The advantages of such a descriptive and mediative approach to contemporary agricultural controversy are more evident when we compare it to other modes of intervention. There are three common modes of intervention, all of which might entail description yet put it in service to other ends. These may be dubbed "answering," "advocating," and "accepting" (see Figure 1.1).

In "answering," we would be conceiving of our intervention as coming to the truth of the matter, as providing correct answers to the questions about which there is disagreement. As good problem-solvers we would describe to ensure that our answers take into account the facts of the situation. As "advocates," on the other hand, we would be conceiving of ourselves as participating, on the side of right, in a political struggle. Descriptions of the true seriousness of the problems would stir others to

Figure. 1.1   Responses to Controversy

Response	Intended Outcome	Limited by
Answering factual questions	objective truth	– which questions are asked – what is counted as a sound answer – uncertainty of the future – lack of information
Advocating policies	the "right" policies	– lack of consensus on the problems – lack of consensus on what counts as solution – counteradvocacy – inconsistent policy outcomes
Accepting the course of dispute	historical witness	– outcomes may not be acceptable or even tolerable – leaves field open to likely opponents
Describing deep differences	true dialogue, climate for accommodation	– makes no substantive prescriptions – makes no assessments of contested "facts" – conflicts that are unresolvable, necessary, or healthy – unwillingness to participate in good faith

join us. Finally, we might find it necessary to be "accepters" of controversies, holding the view perhaps that conflicts in agriculture (for example, between farmers and consumers) are inherent and unavoidable, or that we are powerless to resolve them, yet still finding it important to stand witness before these events so that in future people might understand better and act more wisely.

Our experience in this project has fostered a conviction that answering, advocating, and accepting are not adequate responses to contemporary agricultural controversy. Each approach is subject to trenchant criticisms from the others and each makes assumptions that others reject. The feasibility of the first approach, answering, requires in the first place that opponents agree on the questions to be asked and on what count as adequate answers. Furthermore, for it to work we must assume that we can discover the best agricultural policy to secure the goals we share. To be sure a policy *is* the best we must have the ability to predict the future fairly accurately. Whether we can do this is, of course, not clear. This approach also tends to postpone important policy decisions until major questions have been answered. But, say some, if we put off decisions until we are well informed, we may be committing ourselves to an unwise policy in the present. We do not have the luxury to wait for better information—by the time the potential damage of soil depletion, deforestation, or loss of family farms is fully recognized, it may already be irreparable. Seeking answers in the future is no substitute for action now.

Yet hoping to resolve agricultural controversy by advocacy is also naive. Without consensus on what the problems are and what are the criteria for solutions, victories are likely to be temporary and partial. Advocacy may shape an agenda and secure a hearing, a rider to a law, or a small slice of public support; yet without substantial change in opposing views, continual pressure will be needed just to hold past gains. Indeed, it could be said that current agricultural controversy reflects overreliance on advocacy. Many groups have affected the policy agenda, yet controversy continues unabated. Agricultural policy, it is often suggested, has become a collection of conflicting and ineffective programs. There is, of course, the possibility that some particular advocacy group might finally overcome all rivals. Conceivably, champions of small farming or of corporate farming may someday prevail absolutely. Yet winning doesn't prove policy is sound. The winner's policy may not be the most legitimate, or the most just, or the best in any sense (Schnaiberg 1983).

Simply accepting the situation is also risky. Agricultural policy may always represent a jumble of competing interests, varied circumstances, accidents, and mistakes. .It may be shaped by so many and such diverse forces as to make it unlikely that we will ever be able to come to durable and widely satisfactory policies. Perhaps controversy must evolve in certain ways. Yet we can not be sure that its evolution is predestined to be good. To fail to advocate on the grounds that it will make no difference leaves more room for others to advocate more successfully. To give up on answering questions, to sit back and watch policy unfold, will doubtless lead to some sort of food production system, but there is no reason to think it will in any sense be an efficient, just, or otherwise satisfactory one.

In short, in recent agricultural controversies, none of these three responses—answering, advocating, and accepting—has seemed to work. If answers could end controversy, most of the controversies would be over. Blue-ribbon commissions as well as individual agronomists, economists, and others have insisted repeatedly that they have the answers (Cochrane 1982: 6; Cochrane 1963). But either they are not the answers to the questions others want answered, or they do not meet others' criteria for adequate answers. Nor can righteous advocacy resolve the controversies; the political arena is full of righteous, sincere, dedicated citizens passionately disagreeing with one another. If we could simply trust that things will be taken care of, controversy would not reveal such passionate, partisan commitments.

How Hard Is It to Build a Bridge?

These then are reasons to think that ordinary modes of intervention will not work very well. But are there reasons to think that descriptive mediation will work any better? In particular, is it true that a fundamental failure of participants to comprehend how others think characterizes contemporary agricultural controversy? If it does, where does that incomprehension arise? Is it intentional, or a result of insensitivity or inattention, or does it lie elsewhere? Finally, how is it that descriptive mediation is to contribute?

Over the years there have been several gatherings for the discussion of various matters of agricultural policy by those representing some of the wide range of outlooks that characterize contemporary agricultural debate. Two important examples are an Iowa State "consultation" between agricultural scientists and clergy, particularly campus ministers at land-grant universities, and a Texas A&M "Farm Summit" of agricultural scientists, policymakers, and representatives of commodity and consumer groups and farm organizations. Both meetings took place in 1978 when the Carter administration was actively seeking to broaden access to agricultural policymaking. Each focused on only a few issues:

the Iowa State meeting concentrated on responsible agricultural development in the third world and on the survival of the family farm in the U.S., the Texas A&M meeting focused on traditional pricing and marketing policy issues.

Despite the restricted foci and the generally favorable conditions of the two meetings, participants in both forums were frustrated at the outcome. Something about the process of discussing and negotiating wasn't working. Their comments are striking. At the Iowa State meeting, one of the outcomes of the discussion, an agenda for "the 'new' agriculture" (a version of "appropriate" or alternative agriculture) offended representatives of the old agriculture. They felt stereotyped as unaware and insensitive, and felt presumed-upon to endorse particular social goals, such as making agriculture more labor intensive. When they said things like "I can't sit here as a member of an LGU [Land Grant University] and allow you to say . . ." they were showing that something fundamental to them had not been taken into account in what was being held out for adoption as a consensus recommendation (Conner and Hessel 1980: 121-7). The civil and conciliatory manner of their protests belied the depth of disagreement.

At the Farm Summit, on the other hand, there was a great expenditure of effort to avoid presumption, at least in some of the task forces. Yet, even though the issues were fewer and the range of interests represented (or even considered) reflected only a small portion of the spectrum of positions on agriculture-related issues, taskforce chairpersons were hard pressed to manage the level of disagreement. Even in the report that came closest to recording the diversity of opinion, on "Nutrition, Product Quality, and Safety," one member felt obliged to add a minority report. The authors of the full report, Timmer and Nesheim, admitted that their discussion had failed to discover guidelines for national policy on the issues: "Such a set of recommendations may be feasible, but this Task Force could not find it. Fundamentally different perspectives and values precipitated useful, frequently even enjoyable and humorous interaction, but those differences precluded resolution of the issues" (Doerr in Conner and Hessel 1980:131; Timmer and Nesheim 1979: 185).

Spokespersons at both meetings used metaphors suggesting a great deal of missed communication. At the Iowa State consultation, Lawrence Doerr noted that "we spoke in tongues," while the chairman of one of the task forces in the Texas meeting spoke of "this Tower of Babel with tens of conflicting voices," and of the need "to translate them into a coherent and effective political language" (Kramer in Timmer and Nesheim 1979: 219).

While diversity of interests and issues certainly characterize the problems encountered, remarks from participants in mixed discussions suggest a central problem that cuts deeper. Not only are the differences between diverse viewpoints great, but the means for addressing them are wanting. Even when quite competent representatives of diverse views are brought together, they may not be able to understand each other very fully, or sometimes even at all. Traditional processes for democratic consensus building presuppose that some basis for reaching agreement exists in the capabilities, culture, and practices of citizens. But if the remarks above are any indication, we must begin to confront the possibility that the basis of agreement does not exist.

The Search for a Process

From within agricultural policy arenas the disturbing prospect that disagreements may be unresolvable is often interpreted either as a matter of conflicting values and interests or as a symptom of inadequate information. There is, however, another possibility. Unresolved differences may be the result of people operating within different and incompatible frameworks of understanding. This possibility does not rule out the others, but it does suggest that no amount of effort to get better answers, to advocate better values, or to harmonize competing interests may be enough to resolve the differences. If people process information, assess options, and set goals in basically divergent ways, then reconciling the inputs to those processes will not necessarily lead people to agree on the outcomes (Schwarz and Thompson 1990; Killingsworth and Palmer 1992).

In some ways the concerns about the quality of communication expressed at the two conferences reflect a divergence of frameworks. Speaking "in tongues" suggests much more than an encounter with unfamiliar concepts, divergent values, or even a strange dialect. It expresses serious doubt whether the overt "speech" of the other side is even to count as genuine language, and hence, whether "coherent" translation is possible. The "Tower of Babel" metaphor may be more generous in crediting a coherence to the other guy's speech, but it is no less alarming in its estimate of the seriousness of the overall problem. In the history of science such alarms have been found to be frequent when "paradigms" shift or are called seriously into question (Kuhn 1970). Perhaps here too they are a sign that the basis of agreement has fractured and that the separate ways people process the world no longer fit together, no longer compare coherently (Beus and Dunlap 1990).

When frameworks clash it is natural to expect a concern for the quality of communication to loom large and it does in the literature of agricultural controversy. There are frequent appeals for trust and effective communication, a manifestation that something about the re-education campaigns participants have been practicing on one another has not worked. For example, there is talk of the "need for increased dialogue" (Carlson, Nead in Conner and Hessel 1980: 123, 133; Vogtmann 1984: 33). One author even speaks of the need for "sensitivity training" of those involved in agricultural controversy (Hadwiger 1980: 53). At the Texas summit, there were numerous allusions to the need for a workable decision-making process satisfactory to all the parties to controversy:

> A fair *process* for deciding on policy directions and governmental interventions in the food sector is desperately needed. Many task force members felt that establishing this process for reaching decisions in the nutrition, product quality, and food safety areas was the most important goal for the entire Farm Summit" (Timmer and Nesheim 1979: 170, ital ours).

Timmer noted

> the issue is basically that of finding a *process* with which all interested parties will be satisfied even if they disagree with particular outcomes of the process. We need a decision-making process not governed by a person's political ambitions or the profit motive, that searches for the general interest and treats all interests fairly. . . . the Task Force was *not* able to come up with a blueprint for such a process, but was able to reject all the readily available alternatives, e.g., the Food and Nutrition Board of the National Academy of Sciences (which, however helpful on scientific issues, cannot resolve policy issues) (202).

Jim Turner of the National Consumer's League added that "whatever process was established must involve *all* interests and be perceived by all as a fair process to have the credibility necessary for success" (202).

Perceptions of mistrust, bad faith, and polarization were also expressed. At the Iowa State meeting Don Hadwiger chided "new agenda" groups for presuming that dialogue with conventional agricultural scientists would fail: "Take a positive approach, and I think you'll find that agricultural scientists are humane people. They have some concerns that haven't been raised yet, such as natural resource conservation" (53).

At the Texas summit the clearest example of an inhospitable atmosphere for dialogue was in the discussion of nutrition and food safety.

> The producer activists . . . felt great concern that consumers . . . were so distrustful of information emanating from the food industry that no meaningful communication could take place. The consumer activists felt that producer activists considered them to be biased and uninformed. Some channel of communication in an atmosphere of trust is urgently needed to avoid further factionalization and fear in the food arena (170).

Part of the solution, Timmer and Nesheim noted, lay in the recognition by the participants that those who disagreed with them could still be sincere and act in good faith: "there is honest disagreement over whether such steps [e.g. stronger labeling laws] are necessary or desirable, and at what cost, but questioning the motives of people who seek such change will only exacerbate the tensions and distance between the consumer movements and producer interests" (178). They also recognized that while politics as usual might produce some compromise policy, it would not necessarily be a wise policy: "if disputes are settled in the political arena, then political solutions will be imposed, no matter who is right" (179). John Kramer, author of the summit report on "Agriculture's Role in Government Decisions," had similar views. He noted that intransigence is often politically counterproductive: "agriculture invariably says 'hell, no!', to pesticide regulation, for example, and digs in its heels without exploring reasonable alternative approaches. As a consequence, it is enjoined and prohibited in a harsher fashion than is necessary" (231).

Collectively these comments circumscribe in a preliminary way what is needed to reconstruct a basis of agreement in the face of clashing frameworks of understanding. What is needed is a positive approach that builds trust and is rooted in respect for opponents' good faith. Genuine dialogue and fair treatment among all interested parties are crucial; all must be satisfied with the process even if some disagree with particular outcomes. The process must have an integrity of its own that resists control by external ambitions. Such a process must overcome intransigence, must have means to avoid degeneration into easy presumptions of bias or ignorance and to surmount the facile and futile questioning of motives. In short, such a process must avoid aggravating tensions or fears and actively depolarize factions. Also noteworthy is the suggestion that both politics-as-usual and scientific institutions do not meet these criteria.

The Need to Explore Deep Differences

That these kinds of instincts and perceptions exist among those involved in setting the direction of agricultural policy is heartening. We believe these appeals are sincere and represent an important and growing movement from within the agricultural establishment to accommodate its critics. While some may doubt that such conciliatory gestures are truly in good faith, there is something to be gained by assuming that they are. Dismissing them out of hand before fruitful conversation has barely begun reinforces polarization, but taking them to heart opens up possibilities of reconciliation and constructive change. By the way we choose to act, we shape the conditions of policymaking.

It should not be expected however, that by themselves good faith and willingness to communicate will produce fruitful discussion. For that we need to examine our differences and to explore diverging frameworks to see where they may yet be brought into fuller converse. In both the Iowa and Texas forums participants agreed on the need for dialogue and depolarization, but they did not, as some of them recognized, achieve it. In both meetings there was a groping for words, concepts, and techniques that *could* bridge differences.

Some participants even recognized a need to work with conflicting frameworks. At the Iowa meeting, Jerry Stockdale, a social scientist, spoke of the existence of "a set of myths, biases, and presuppositions which hindered our ability to analyze political-economic and environmental realities." He argued that the "assumptions out of which we assess policy alternatives" had to be made explicit. He recognized that it was "not possible, nor necessarily desirable, to eliminate all bias, myth, and ideology; but individuals should attempt to understand their biases and, where appropriate, attempt to minimize their impact" (73).

The myths and biases Stockdale was referring to were those of the agricultural establishment, which had led, he felt, to undesired social changes in less developed countries. Yet such calls often have been as much a tactic in controversy as a recognition of what underlies deep disagreement—it has been only the *others* who are seen to possess the troublesome "myths" and "biases." The request to see the controversy in these terms has thus often been a request to surrender to another's outlook, an invitation to "be realistic, think like me."

How much Stockdale's suggestions were perceived in this way is not clear, but they did serve as a jumping-off point for the group at the Iowa meeting that focused on curriculum reform. Lawrence Doerr reported the group's recognition of the

> need to build understanding, awareness, and to foster conscious choices about the values, the meanings, the perceptions, and the political and social biases by which we live with one another in this global village. We must learn something more than appreciation, something more than tolerance of the differing perceptions of reality and the differing technologies which exist side by side and contribute in the broadest sense to our basic interdependence (131).

The statement—Hadwiger called it "a committee poem"—points to the need to see how values, meanings, and perceptions go together with social action and technology. By calling for something more than appreciation or tolerance, Doerr was recognizing the need to represent "the differing perceptions of reality, existing side by side . . . which contribute to the interdependent settings in which we live" (131). In the context of Stockdale's talk, the implication here is clear that well intentioned programs and policies can avoid inadvertent damage only by an active effort to explore, understand and build bridges between "the differing perceptions of reality" by which we live.

Since the Iowa and Texas conferences, problems of polarization in agricultural policy making have been evident in a great variety of settings, and many efforts have been made, both from within the agricultural establishment and from without, to generate better policymaking processes. Perhaps all of these efforts have been well-intended, yet successes have been few and transitory.

What Description Might Do:
Description as an Antidote to Presumption

A central part of the communication problem in both these forums was presumption. People didn't realize how deep their disagreements were; they assumed—naturally enough, as we all do—that what made sense to them would make sense to others, who perhaps simply didn't have sufficient information or hadn't thought things through carefully enough. This is characteristic. In deep disagreement people presume most when they understand least. They may not see where opponents are coming from or how they can hold such "outrageous" views with good intentions. They may not be aware of their own assumptions or fully understand the implications of their assumptions. They may not see how easily one can be trapped by one's own reaction to controversy.

The descriptions of ideologies of agriculture we give below have been undertaken with two aims. The first is to facilitate communication. Opponents in controversy often fail to communicate effectively. In part this failure reflects differences in idiom—the idiom of the poet Gary Snyder is not readily translatable into the idiom of the agricultural economist Luther Tweeten (Snyder 1984; Tweeten 1983). The failure is also due to rhetoric. Many writers in agricultural controversy are trying to bolster, celebrate, or advocate their own position. In doing so they stigmatize, totalize, and insult the opposition (Coleman 1982: 382-3). That many do so even without trying to shows how deeply seated the communication problem is. When these problems occur, careful and balanced description can alert people to dangerous presumptions and illustrate how to avoid them.

The second aim of our descriptions is to create a climate for limited compromise, mutual accommodation, and increased tolerance in controversy. It is often possible to work with disagreements constructively. If people look beyond offending presumptions they may find room for limited compromises. In some cases they may find no real opposition. In others there may still be reason to tolerate one another's activities and agree to disagree. Description of deeper differences helps to realize such a climate by clarifying what underlies the positions taken, how views are transformed into arguments for policy prescriptions, and hence what room there may be for compromise.

The upshot of these objectives is a commitment to work toward a political climate more favorable to open discussion. But that does not imply that conflicts will go away. Some differences may be irreconcilable. Some disputants may not want to try to talk things out. Some conflict may be healthy or necessary. In seeking a larger place for open discussion in policy processes, we accept these limitations. But where exactly do they fall? We want to extend the benefit of doubt as widely as possible. If there are ways to transform disagreements from obstacles into resources, then we must seek them actively with respect for the motives and intentions of all the participants in controversy. And we must look out for and be ready to set aside our impulses to write off some of the other fellow's views. If matters really are intractable, we will find it out by our failure to make progress with the dispute. But if the dispute is workable, we will discover that only by trying to work with it.

To encourage discussion, description must be undertaken with two points constantly in mind (Figure 1.2). First, it needs to be impartial. If the description appears to any of the participants in agricultural controversy to be loaded against their views or in favor of others', it will not achieve its aim. There are ways of cultivating impartiality—for example, by maintaining the focus on facilitation and avoiding the temptation to answer or advocate; by stressing the coherence and rationality of each viewpoint; and by addressing the same questions to each. Second, the description must be in depth. Merely repeating each outlook in its original idiom is not effective. Instead, description should translate—if feasible into a common idiom. Where that is beyond reach, a series of translations can still be given, one by one, for the other views. Either way there will be enough of culture and human experience shared in common to permit more fruitful communication (Geertz 1979).

Events since we began this project suggest that favorable changes in the political climate for agriculture have already begun. New journals have been started to discuss agricultural issues; new associations have brought people together from different backgrounds; the issues of agricultural controversy are being addressed in the curricula of agricultural colleges and universities; and new research programs are focusing on neglected areas and are exploring a range of ways to reshape agriculture.

Figure 1.2   Conditions of Adequate Description

Condition	Meaning	Means
Impartial	not perceived as loaded against the views of any disputants	– focusing on facilitating communication, not answering or advocating – stressing the coherence and rationality of each view – addressing the same questions to each view
In Depth	explores the meaning of each outlook's preferred language	– translating into a common idiom – translating seriatim for each opposing view

In this ferment themes such as sustainability, justice, health, family, community, and environment are finding a new place in the discussion.

But the basis for agreement on the direction and means of change is in many ways no clearer today that it was over a decade ago. Nor is it likely to become so of its own accord. If those concerned with U.S. agriculture are to shape its future in fair, sound, and balanced ways, the sources of factionalism and the discontinuities between different ways of understanding agriculture must be addressed more actively and more effectively.

This book proceeds in the following way. In the next four chapters we dig down to locate the "deep disagreements." We ask not *why* people disagree, but *how* they manage to disagree, given that they are living within a single culture, using a common language, and sharing many conventions about methods of coming to and defending positions. Is their disagreement, as many believe, principally characterized by factual errors or mistaken values? Or, at a slightly deeper level, are they appealing to different contexts of discourse, something like different academic disciplines, to find their answers? Neither of these satisfies us; neither allows us to understand what is coherent, what makes sense to the participants, even what makes it necessary for them to say and do what they do. To find this sense we search at a deeper level—the level of ideology. In Part Two we describe in depth three ideologies prominent in U.S. agricultural controversy. Part Three focuses on the appearance of ideology and ideological conflict at two important sites of agricultural policymaking: the legislative hearing and the formulation of agendas for agricultural research.

Notes

1. The works of Schwarz and Thompson, 1990; Norton, 1991; Beus and Dunlap, 1990; and Wojcik, 1989 reflect a similar concern for communication and bridge-building but employ a more traditional authorial role.

2

Sources of Polarization: Facts vs Values

Since we want to open up a serious exploration of deep differences in agricultural controversy, we must be careful not to reinforce the patterns of polarization that already exist. Some sources of polarization do, of course, really lie in the content of deep disagreements, but a great deal of tension appears also to arise from the way we organize and present our contributions. What this means is that the good intentions of participants may not be enough; the arenas of agricultural policymaking are full of well-intentioned participants who all too often have aggravated tensions inadvertently and unknowingly. There is much to learn from their mistakes.

The complexity of policymaking itself can promote polarization. Underestimating that complexity or hoping to escape it, participants take shortcuts. At the heart of these shortcuts are presumptions either of the value of one's facts, or alternatively of the facticity of one's values. Both critics and defenders of established agriculture often posit a fundamental distinction between "facts" and "values" and privilege one side of this distinction in proposing policies. Often, though by no means always, it is critics who champion "values" and defenders who champion "facts." Moreover, each side presents its preferred realm as insulated from the other: either we are enjoined to adopt better values with little regard for empirical evidence of how the world of agriculture actually works, or we are exhorted to accept "the facts" as somehow sufficient to resolve what is the right thing to do. Often the collision between facts champions and values champions leads to mutual frustration and irritation; dispute escalates, civility breaks down.

In this chapter we argue that to separate the realms of facts and values is not a suitable approach to agricultural controversy: to try to do so is to oversimplify both. Our aim is not to favor one side or another but to discover how to avoid organizing discourse in ways that inflame controversy.

The Temptations of Facts-Values

Policy discussion usually involves several seemingly different kinds of pronouncements. To arrive at appropriate policies we normally do a variety of things: we review the results of past policies, predict the probable consequences of alternative policies, assess the benefits, costs and risks of prospective options, compare the worth of different options, and decide on and recommend one. In this spread of activities, we are sometimes asking what has happened, or what will happen under certain circumstances (and stating our answers as "facts" or as predictions); sometimes we are evaluating the merits of various outcomes, or the effectiveness of various means (and expressing our conclusions as "value judgments"). Sometimes we are telling others what we think they should or ought to do, or we are making recommendations for further consideration (and stating them as suggestions or prescriptions). And sometimes we are making decisions or declaring the results of a decision taken (and expressing them as imperatives or declarations).

It would seem that we could separate these pronouncements into two groups (Figure 2.1): those in which we assess the worth of something or prescribe or recommend a choice ("normative" pronouncements) and those in which we are *merely* stating the way things are, have been, or will be ("factual" pronouncements). The first group of "normative" pronouncements all seem to involve an evaluative element, one that at first sight appears absent from the second group of "factual" pronouncements. Moreover, the evaluative element involves a characteristic problem. As the value theorist Kurt Baier has observed, in valuing things people may disagree not only about what traits they have but also "about the rightness of the criteria employed in ascertaining the 'value' or 'goodness' of the thing in question" (1958: 75). Critics and defenders of conventional farming practices, for example, may differ not only on the consequences or effectiveness of a certain practice, but also on the criteria for judging its worth. Defenders might point to high yields as indicating the value of the practice, while critics may object to deleterious environmental consequences. Thus the dispute involves differences over what "good farming practice" does or should mean. The question of what criteria we should use in judging something to be good farming practice becomes part of the issue whether the practice *is* good farming practice. When the issue is complicated in this way it would seem to call

Figure 2.1 Factual vs Normative Pronouncements

Type of Statement	Range	Issues Raised
Factual	– descriptions of past and present – predictions of the future – predictions of consequences of prospective actions	– do things actually have the features described or predicted?
Normative	– assessments of costs, benefits, and risks – comparisons of the worth of different outcomes or the effectiveness of different means – suggestions, recommendations, or prescriptions of what to do – declarations of decisions taken	– do things actually have the ascribed features? – is the value of those features judged appropriately, i.e., according to the right criteria of worth?

for different management because more is involved than just using agreed upon or conventional criteria to check the truth, correctness, or accuracy of what is being claimed.[1]

To separate issues of fact from issues of value may thus seem a wise course. In policy discussions, where both normative and factual pronouncements are common, it is often expected that discussion will proceed more smoothly and clearly if those issues are separated out which we can hope to resolve through experiment or observation. Then, with the facts before us, we can battle over how to evaluate options. When all sides have a common view of the meaning of a particular issue within the larger controversy, and therefore agree on how to determine what is the case on that issue, some contentiousness may be avoided and remaining issues may become more manageable. Over the years, a fact-value distinction has become standard in many quarters, including agricultural policy discussion. If applied with circumspection, the distinction may still be of some use in organizing policy discourse. But many problems beset this mode of organization, and as we shall shortly see, in the hands of strident combatants it becomes too easily and too often a source of polarization. Let us take a closer look at how it is used in practice and then go on to consider its inadequacies.

Strident Combatants

Critics of conventional agriculture have often focused their criticisms on the values it appears to represent. Problems of practice—from soil erosion to overproduction, from pesticide residues to increasing concentration of control of resources—are presumed to reflect mistaken judgments or faulty presuppositions about what is important or about what criteria should be used in judging worth. If the problem is wrong values it seems natural to think the solution will be right values: replace mistaken judgments by correct ones, or faulty presuppositions by sound ones, and right practice will follow of its own accord.

One critic, Richard Conviser, opens a discussion on "Appropriate Agriculture" thus:

> The crisis of modern large-scale agriculture is rooted in the crises of modern science and economics, particularly in their disregard for the wholeness of nature. Scientists have substituted for the meaningful whole that is nature an analytical, fragmented, objectified, constructed reality. Economists have not respected the wisdom of natural systems but have instead focused upon agricultural "inputs" such as land and labor as commodities to be manipulated and exploited in a human domination of nature. Agricultural practices which have developed out of these ideologies produce high short-term yields at the cost of longer-term consequences devastating for both humans and nature.
>
> For agricultural practices to change substantially, values must emerge which are alternative to efficiency and short-term productivity (1982: 436).

The passage begins with a list of errors: the "disregard" of "wholeness," the substitution of an artificial view of reality, the failure to respect "the wisdom of natural systems," the attempt to dominate nature. It then explains that these erroneous values or "ideologies" are having "devastating" practical "consequences." New values will produce "appropriate" agricultural practices.

Conviser's argument is based not only on the claim that practice is value-driven, but that the values that should drive it are so obvious as to be beyond the realm of criticism. That those values are right he represents as fact. Not only are they presented as clearly the right values, but they are right because they are in accord with nature. The power of Conviser's argument depends on our willingness to read claims about "wholeness" and "wisdom" and even the wrongness of "human domination of nature" as givens—not just as statements of Conviser's own perceptions or values. The wrongness of conventional agriculture then, lies in its failure to adopt the right values.

This presupposition of Conviser's—that the right values are given by nature—makes it seem unnecessary for him to defend them. He does discuss criteria of "appropriateness" in the "appropriate agriculture" that must be developed: "appropriate" technologies are to be "small, simple, capital cheap and non-violent." But his argument that conventional practices "do not fare well against these AT values" (436) belabors the obvious at the expense of ignoring the deeper issue: why AT values are the right ones by which to judge conventional practices.[2]

Conviser's stance is polarizing precisely because it evades the central issue of the appropriateness of his basis of judgment. Far from winning converts, his failure even to acknowledge alternative criteria as tenable is likely to reap enmity, especially from agriculturalists who have had an earful of this sort of thing more than once.

Agriculture's critics have not had a monopoly on privileging values. An argument frequently raised against the viability of organic agriculture is that it is impossible (or simply wrong) to change the values consumers express in their shopping decisions: a greater concern with price, appearance, and convenience than with flavor or nutrition. These too are touted as values that must be observed in organizing our agriculture (see also chapter 12 below). In granting no access for criticism this stance can be quite as polarizing as Conviser's.

Privileging facts, particularly scientific facts, can also be polarizing. Even if we assume the facts are truly facts, questions still arise of their significance. Are the facts relevant? and how? Are they complete? What conclusions do they warrant? Do they represent problems needing solution or constraints to be lived with? The development of resistance to pesticides by pests, for example, may be seen either as a problem to be solved with more powerful pest control techniques or as an unavoidable consequence of biological law, in which case it may be wiser to practice mixed cropping to diminish pest damage.

The most familiar (and probably the most polarizing) examples of privileging facts come from public sector agricultural scientists responding to their critics. In extolling the accomplishments of agricultural science, some of its practitioners have insisted on the value neutrality of scientific method. As practitioners of that method—the people with the unique access to the facts—they have represented themselves as the exclusive arbiters of truth and reason in matters of food and agriculture.

In the early 1970s scientists from a broad range of agricultural disciplines organized the Council for Agricultural Science and Technology (CAST) to inform the public of the role and content of agricultural science, to correct "misinformation," and to rebut agriculture's critics (Black in CAST 1972: 1; Hutchcroft 1982; Baker 1980). CAST publications have included both spirited defenses of agricultural science and ostensibly value-neutral summaries of scientific information for the general public.

A common assumption of early CAST publications is that facts speak for themselves. In a CAST occasional paper, Theodore Hutchcroft described the role of a special CAST task force on the risks of nitrite in meats in the national controversy over USDA and FDA regulatory policies.

> The people in Washington who fought this battle have said the CAST report was one of their most effective weapons. CAST, of course, did not take a "position" for or against the USDA/FDA proposals. It merely made it possible for knowledgeable scientists to assemble the research facts so they could be an effective part of the decision-making process (1982: 2).

Here the CAST report was both a key "weapon" in a "battle" and yet still neutral. (Indeed it was precisely the assertion of neutrality that gave it much of its power.) Not only does such language cover up possibilities of scientific uncertainty and error (especially important in this case where there was conflicting evidence and divergent scientific opinion), it presumes that facts imply policies: knowing the facts is presented as a shortcut past the complex questions of policy determination.

This sort of stance sometimes takes a form that echoes Conviser's failure to justify his standards for judging agriculture. R. H. White-Stevens, for example, in a 1972 CAST publication on public information strategies, offered the following "axiom" for public consumption: "all foods are 'natural' and, with the exception of table salt, 'organic'; ... to charge double prices for so-called 'natural organic' foods is a shabby fraud on a gullible and unsuspecting public" (CAST 1972: 6). Just as Conviser treated AT values as given, White-Stevens insisted that particular definitions of the terms "natural" and "organic" were the only correct ones. Because those meanings were presumed to be grounded in nature or science (rather than in particular cultures or idioms), they seemed to escape the need for defense. Since the definitions were "merely" statements of fact (like Conviser's AT values) we were enjoined to use them to determine the truth of statements about "organic" or "natural" foods. By privileging one set of definitions, White-Stevens bypassed all the questions that those using other definitions were attempting to raise.[3]

This too is a common shortcut. Agricultural controversy frequently involves conflicting uses of key terms like "crisis," "progress," "efficient," "productive," "sustainable," "healthy," "safe," and "nutritious." But to dismiss competing uses without offering or even admitting the need to justify one's own does nothing to resolve differences. It only antagonizes the opposition and encourages them to find nefarious motives behind such seemingly wilful obtuseness.

It is a matter of no little irony that the fact-value distinction within which both defenders and critics have elected to organize their discourse has aggravated their differences. Too often the distinction has been an opportunity to employ rhetorical techniques that, it is hoped, will bypass the controversy itself. On the one hand some have treated their values as facts, on the other some have insisted that the value of their facts was obvious. Both have ignored the need to justify the criteria they used either to judge value or assess fact. If the point of the distinction is to separate cases where the criteria of judgment are at issue from those where they are not, then the distinction has not worked. Neither the privilegers of facts nor the privilegers of values have been willing to allow that the contested issues include any elements of their own position, yet both insist that elements of the opponent's position are at issue.

It may seem that this concern for justification is only a philosopher's issue; philosophers, after all, are famous for never being satisfied with the justifications used in ordinary life. But here failure to recognize a need for justification spawns a host of deficiencies that often obstruct policy deliberation. We need to consider what level of justification is needed and how the provision of that justification can move us beyond the paralysis produced by competing attempts to separate and insulate the realms of facts and values.

Normative Oversimplification

In part what we mean when we say value statements like Conviser's are unjustified is that they are vague. Justification is not simply a matter of appealing to some prior metaphysical assumptions or some ultimate ethic, it is also a matter of showing how values are (or might be) related to facts and actions. When criteria of evaluation are left extremely vague, there is no way to tell which specific measures are to be favored and which are not, or which kinds of flexibility in practice are appropriate. Questions of what is technically, economically, socially, or politically feasible are obscured; and the complex ways in which the bases of evaluation and the circumstances of action inform and guide practice are ignored (see also Willhelm 1967).

C. Dean Freudenberger's 1982 essay, "The Ethical Foundations of the Idea of Agricultural Sustainability," illustrates these problems in its attempt to "re-formulate a normative concept of agriculture":

GOOD AGRICULTURE IS THAT WHICH CONTRIBUTES TO JUSTICE,
PARTICIPATION IN THE ON-GOING PROCESS OF CREATION, AND
IS SUSTAINABLE BOTH NOW AND INTO THE DISTANT FUTURE.
. . . One must ask in view of the normative construct: "Is there anything
that we do today in agriculture that comes close to these norms and
goals?" (630)

In asking for justification, one is in part asking for explanation. One
might ask what Freudenberger's notion of justice is (and why), what is
meant by participating in the "on-going process of creation" (and why
that is important), and what is meant by sustainability. One might also
want to know how to balance the three goals. Freudenberger tries to
answer many of these questions. He qualifies the scope of his prescription (626) and urges "a sense of proportion" (627) in applying it to
situations where social justice is at issue.

Yet he still does not escape the problems of vagueness and oversimplification, even in the case of sustainability, which may seem the most
concrete of the three goals. Freudenberger explains that

> An ecologically sustainable agriculture is . . . one where the notion of
> permanent carrying capacity is maintained, where yields (agriculture,
> forestry, . . . fisheries, water use) are measured on a sustainable basis
> rather than on the conventional criteria of the maximization of yields per
> acre, profit from investment, and where waste products can be absorbed
> back into the eco-system without damage (626).

To those who do not agree with Freudenberger (and even perhaps to
some of those who share in general his criticism of conventional
agriculture), this explanation will not resolve the problem of vagueness.
What counts as maintaining "the notion of permanent carrying capacity?"
What does it mean to have a "sustainable basis" for measuring yields and
how can one tell if it is actually used or not? What does it mean and
how can one tell whether waste products are absorbed "without
damage?"

All participants in contemporary agricultural controversy can mount
a claim to be endorsing sustainable agriculture, though they don't always
find it rhetorically useful to do so (Douglass 1984). Hence all can make
a claim to be meeting each of Freudenberger's criteria (Hamlin 1992; see
also Chapter 6 below). In determining whether a given agriculture meets
the criteria of sustainability they will be making different assessments of
the likelihood of social, personal, or political changes (e.g., can the
growth of human population be controlled, and by what means?) and of
the capabilities of future technology (e.g., will we find satisfactory ways
of combating the salinization that comes from irrigation?).

A succinct way of capturing the vagueness of Freudenberger's prescription is to say that he has leapt from a general premise ("Good agriculture is sustainable") to a specific conclusion ("We should practice sustainable agriculture"). Even if one accepts his premise and restricts its scope (as he seems to intend) to judgments about the agricultural use of non-renewable resources, the desirability of material growth vs social justice or spiritual growth, or the consequences of trends toward bigness (626), the outlines of a sustainable agriculture that is technically feasible, financially prudent, socially equitable, or politically acceptable are not clear. Nor is it clear how the value of sustainable agriculture is to be weighed against values for agriculture that others might advocate, values such as a healthy agricultural economy, economic growth, adequate supplies of food and their just distribution, stable rural communities, or low food prices for consumers.

Like many critics, Freudenberger here takes the view that we need first to decide on the right goals. His assumption is that nearly anything is possible: "Science and technology can provide us with the means to create almost anything that we want" (628). Through this assumption problems of means, of implementation, and of the inertia of political, social, and natural circumstances—problems that are typically at the heart of defenses of conventional agriculture—are trivialized. Some sorts of sustainable agricultures may well be both feasible and desirable, but Freudenberger's technological optimism doesn't bring us closer to recognizing them. Rather, it obscures the connections between desirability and

Figure 2.2   Competing Oversimplifications

Type	Typical Effects
Normative— insulating values	– values are left too vague – no criteria for choosing specific measures – some relevant facts are ignored – questions of feasibility are obscured – circumstances of action are ignored – means and ends are not reconciled
Factual— insulating facts	– the values presupposed by an inquiry are ignored – the meanings of facts are assumed – questions of relevance are begged – questions of completeness are begged – criteria for judging worth are left out – the value of science is put beyond discussion – other ways of judging facts are demeaned – science is assumed to be value neutral

feasibility and takes for granted that if we straighten out our "values" solutions will follow. The need to reconcile means with ends is not acknowledged (see Figure 2.2).

Epistemic Oversimplification

Just as appeals to values founder on the issue of what is to be the relation of the values to the world of facts, so appeals to fact founder on the question of what is to be the relation of facts to values. Claims to fact often mask evaluative questions, for example, of what value judgments are reflected in the creation of a particular claim to fact or of whether a set of facts is adequate to or even relevant to the purposes at hand.

Strident defenses of conventional agriculture are often underlain by assumptions about the nature and worth of science that are rarely made explicit and even more rarely critically examined by agricultural scientists. In a list of twelve axioms to impress on public consciousness R. H. White-Stevens offers the following:

> That the scientific method owes no allegiance to any political party, religious credo, ethical philosophy or material power; that it is simply a method involving four steps:
>
> a) Observation
> b) Deduction
> c) Experiment
> d) Induction or Prediction
>
> These steps are totally objective, replicable, calculable and reliable; what is done with them may have ethical, political, or religious intent and overtone, but not the method itself (CAST 1972: 6-7).

White-Stevens expresses the common view that there is a scientific method that transcends other ways of knowing: it alone produces truly reliable facts. Science, then, is the ultimate voice of authority, at least in matters of fact.

But like Freudenberger's, White-Stevens' approach leaves us without a way of answering many of the key questions of agricultural controversy. A host of problems of scientific methodology itself and of its application to agriculture and similar matters of policy have been raised in recent decades both by philosophers and sociologists of science and by critics of conventional agriculture.[4] The standard policy question—what research objectives should agricultural scientists pursue—of course looms large. But more subtle issues of methodology follow in its wake that are of critical concern to many different groups: What observational or experimental languages should agricultural scientists use? When are observations or experiments valid? What conclusion can rightly be drawn from valid observations or experiments? White-Stevens speaks of "induction or prediction" but says nothing about how we are to shape our inductions on the basis of experiment and observation. When is an hypothesis confirmed by the data and what level of confirmation is enough to warrant accepting the hypothesis? Confirmation of a general hypothesis does not automatically guarantee its truth; other data may raise doubts, other hypotheses may account better for present data. And even if, as falsificationists, scientists agree that the claims they hold are only tentative—merely hypotheses that have not yet succumbed to attempts to falsify them—they will still face the question whether falsification itself is negotiable. After all, in a pinch, we can always presume some kind of auxiliary hypothesis to account for the failure of our observations to match our predictions. The history of science is loaded with examples—perhaps phlogiston has a negative weight, perhaps the ether lacks resistance, perhaps some undiscovered heavenly body is distorting the orbits of the outer planets. Sometimes such seemingly ad hoc evasions of falsification even turn out to be right, as in the eventual discovery of the planet Pluto.

Indeed, there is no fixed rule in science to decide when we must abandon an hypothesis, or for that matter when we have enough evidence to decide that a conclusion, even a tentative conclusion, is warranted. In 1953 the philosopher of science Richard Rudner pointed out that there are no patent answers to this "when are we done" question. In accepting a generalization a scientist is making a judgment that the evidence is strong enough to warrant acceptance. If the judgment is being made rationally, the scientist will take into account the significance of being mistaken and will demand stronger evidence when the risks are greater. An engineer, for example, will demand more proof of the safety of a major flood-control dam than a small farm-pond dam.

Cases like this abound in farming and agricultural policy where we are frequently seeking scientific warrant to make major and risky environmental and social changes. They make it clear that however ideal the scientific method may be in the abstract, those "ethical, political, or religious intent[s] and overtone[s]" that arise in its application are pervasive and inescapable. In Rudner's view, it is incumbent on scientists to articulate the judgments of importance that they must make in deciding whether to accept a generalization on the basis of available evidence.

Rival scientific claims in agriculture often reveal several different problems in coming to closure. Take for example the controversy over "organic" vs "chemical" systems of tillage and fertilization. In field tests comparing organic methods with conventional methods, organic methods are often substantially less productive. But are the comparisons fair? Advocates of organic methods, who place much emphasis on both the physical structure and the microbiology of the soil, may claim that trials on plots previously used for tests of conventional pesticides and fertilizers are meaningless, since many years may be required for soil life and soil structure to recover from conventional agronomic practices (e.g., Walters and Fenzau 1979: 167-8).

Longer tests or more exhaustive tests won't necessarily help: the argument could go on in this fashion indefinitely, without warranting the conclusion that one side or the other was wilfully ignoring the facts. And even if there were finally agreement on what yields can be obtained with various agronomic practices, there might well be further disagreement about what policies those results indicated. There might be compelling reasons for adopting organic agriculture even if a sacrifice in production (in terms of bushels per acre) were ultimately required.

It might be argued, for example, that quality is the real issue and that organic produce has superior quality (Allaby and Allen 1974: 72). Quantitative comparisons of yield, then, would be misleading; the products of one form of production would be fundamentally different from the products of another. But surely this too could be put to test, one might think. Yet our tests of differences and similarities will depend on what variables we hold to be significant and on the maturity of our science: the list we bring with us of what differences in the constitution of foods are possible and how they can be demonstrated with the analytical techniques at our disposal. Philosophers of science warn us to recognize our knowledge as provisional; we can not logically prove that there are *no* significant differences, only that we could find none of the differences we looked for using the best techniques available.

Even if no differences are discovered there may still be reasons to hypothesize the existence of differences currently beyond the reach of analysis. To some this may seem the last refuge of someone who is losing an argument, a stubborn and antisocial unwillingness to accept facts he or she doesn't like. Yet it can also be seen as an invitation to explore a new level of natural complexity, a demand that science must progress and scientists must not complacently tout old maxims or dogmatically rely on old methods. There are many cases in the history of science, for example, the science of nutrition, in which the positing of unidentified trace factors turned out to be a good thing to do. Prior to the development of the trace nutrients concept at the beginning of this century and the subsequent discovery of vitamins, most scientists saw nutrition as a matter of supplying the body with building stuff (proteins) and energy stuff (fat and carbohydrates). Not only were what we now recognize as deficiency diseases not connected with trace nutrients, but they were not always recognized as discrete clinical entities (Ihde 1964: 644-56). The recognition of a common cause (an absent nutrient) helped scientists recognize links among symptoms formerly thought to be unrelated. Expectations tell us what to look for and where to look; we cannot pronounce on the non-existence of what we have not yet imagined or articulated; we must acknowledge that our pronouncements on similarities or differences only summarize the tests we have carried out. Even if no qualitative differences are expected in the produce grown under different agronomic systems, there may still be differences in deciding what a measured rate of production means. Some may argue that the proper measure is not production per acre, but sustainable production per acre: one must average production from many growing seasons to see how sustainable it is. Practices that deplete topsoil may lead to declining yields, while a form of husbandry that initially yields less might be more sustainable and hence yield more over the long run.

Determining what facts about yields mean is thus fraught with tension, even when there is agreement on what the facts are. To assess their significance requires evaluative judgments, which in turn reflect assumptions on a wide range of issues: the social functions of agricultural science, the sorts of food and systems of production and distribution that people will accept and use, the limits of future technologies.

Those who favor conventional agriculture will also find in their outlooks good reasons for assessing facts in the ways they do. They may argue that the farmer's financial solvency should be pre-eminent: it might be true that after a lengthy conversion period organic methods would be as productive as conventional, but if farmers go broke in the interim that is irrelevant. They may admit that organic and conventional produce might differ in some unique and obscure way, but still hold that the immediate concerns of consumers (for low price, for appearance and convenience) do not warrant devoting resources to exploring that possibility. They may also recognize that various high-yield practices will stress or deplete a farm's resource base. Realizing that such practices may engender serious problems for farmers and agricultural scientists of the future, they may still recommend their use because they permit today's farmers to compete in world markets and to feed a hungry world efficiently.

Finally, since agricultural science serves social goals, its implications may be construed differently by proponents of different goals. Those seeking to maximize the international competitiveness of U.S. agriculture, for example, might see the high yields of conventional methods as indicating how to structure farming: farms should be large enough to reap the benefits of such input-intensive farming. To others with different social goals, alternative methods that yield less might still make for highly satisfactory ways of farming when used in conjunction with other enterprises and on farms with a unique balance of resources—of labor, equipment, capital, crops and livestock, and market access. In the first view, science would dictate practice and policy, laying out the proper course of action for farmers to follow and policymakers to facilitate. In the latter, scientific knowledge might be seen as providing a set of possibilities, tools for farmers to select as suits their unique farm. Policy might facilitate diversity by ensuring that public institutions respond to farmers' diverse needs.

None of these problems of closure is necessarily a matter of one side being unscientific or irrational, ignoring evidence or wilfully expressing contempt for authority. Quite the contrary, each may be behaving in a way that is highly rational, à la Rudner, in including the context and consequences of experimental work in assessing results. But because each assesses risks and possibilities differently, they reach different conclusions. The authority of science is not in jeopardy in such a view, nor is the utility of scientific knowledge. What does become important is to bring to the level of conscious articulation, reflection, and discussion the judgments and assumptions that the doing of agricultural science involves.

Many scientists recognize all this, just as many non-scientists recognize the complexity of translating values into actions. Yet too often parties to controversy wave the flag of science or claim to speak from the high ground of right values. To attempt to prevail in a controversy by assertions of invulnerable authority is almost always to imply that the controversy is not legitimate: the opposition must be ignorant, unintelligent, or morally bad. As a tactic it is likely to be counterproductive. Accusations like these end conversations; controversy becomes bitter and polarized.

Considerations like these also suggest that the fact-value distinction is unlikely to advance us much in resolving agricultural controversies. Too much remains unsettled: what the right values are, what the implications of those values are, what the feasibility of acting on those values is, what the facts are, what the facts mean, what the risks are of mistaking either facts or values. While there may be need to categorize the different kinds of pronouncement that arise in policy discussions, it is doubtful that enforcing a strict dichotomy between facts and values will be either worthwhile or tenable (Michalos 1983; Schwarz and Thompson 1990: 23). If the aim is to support one's positions with the best reasons available, then more of the underlying complexity will need to be brought out, including especially the complex relations between the bases of evaluation and empirical methods of analysis.

Responsibilities of Scientists and Advocates

Some lessons can be drawn from this examination of strident combatants (see Figure 2.3). For critics of conventional agriculture, particularly those who have argued that it reflects the wrong values, three points loom large in shaping effective advocacy. First is the need to attend to the normative views of one's opponents. Only by taking account of their views can one hope to see what is necessary to justify one's own evaluations and prescriptions, which very likely will utilize criteria, standards, or rules that opponents will not accept in advance. Articulation of one's own normative basis may not compel others to agree, but it will be a means to avoid the bruising of others' sensibilities that comes from slighting or ignoring them.

Second, it is not enough to articulate in general terms what one feels to be important. Citing such concerns as resource sustainability, ecosystem integrity, environmental preservation, distributive justice, interpersonal community, or personal fulfillment may help one focus one's thinking somewhat but such terms do not necessarily mean the same thing to all parties, nor do they bring with them precise practical instructions. Thus, to advocate any of these general values carries with it the burden to make clear what must be done to realize them. In meeting

Figure 2.3 Responsibilities of Scientists and Advocates

Advocates	– take into account the normative views of opponents and how they differ from your own – make clear how the values advocated are to be realized in practice – address proposals to present contexts and circumstances
Scientists	– don't take science for granted—acknowledge unsettled states of science; back claims with evidence not authority; acknowledge the limitations of arguments and conclusions – make explicit the value judgments conclusions presuppose and defend their appropriateness – make explicit the full range of problems of interpretation and questions of relevance, and if expert judgment is preferable, explain why

this burden, advocates face a third responsibility of thinking about change in the context of the present policymaking environment. To acknowledge the enormous inertia represented in existing patterns of farm structure and associated institutions of science, policymaking, finance, marketing, and agricultural inputs will be sobering, but to ignore the actual circumstances in which action must be taken is to risk the abrogation of one's preferred prescriptions: they cannot be the right thing to do if under the actual circumstances they cannot be done at all. Effective advocacy anticipates such objections and qualifies and shapes prescriptions in the light of knowledge of circumstances.

For defenders of conventional agriculture, particularly those who see it as manifesting the application of science to agriculture, it is important to reconsider the nature of science. Taking science for granted doesn't lead to effective defense of agricultural practice if the critics do not take it for granted also. It may well be tempting to appeal to science as a way to escape the political fray. Speaking as an "expert," one seems to get the best of two worlds: the authority of science is made to support a favored policy while the need to defend the policy on normative grounds is obviated. Yet this strategy can undermine public confidence in science, for both sides can play the same game. These days, when experts weigh in on one side of a dispute they can expect to face experts on the other side. The public wonders who to believe and ends up distrusting scientists (and science) altogether. Instead of assuming that science always speaks with one voice, defenders of agricultural science might find it better to acknowledge unsettled states of science and even to own the various kinds and degrees of uncertainty that are normal aspects of settled science. By being more candid about the limitations of scientific certainty, agricultural scientists might gain greater respect.

A second responsibility for agricultural scientists is the need to attend to the suppositions that inevitably accompany making conclusions in applied sciences. The same sorts of questions that we urge on critics also apply here: What suppositions are being made? What is the basis for the conclusions drawn? And why are those suppositions (or that basis) the right ones or appropriate ones to make for the questions at hand? Especially when evaluative assumptions are involved, scientists have much to gain from recognizing their suppositions, making them public, and defending them. They may not convince critics, but at the least they will have helped to clarify the issues and will have avoided the counterproductive result of appearing to have dismissed their opponents' views.

A third responsibility for scientists arises from the necessity to interpret the questions asked of science, and in the process to explain how scientific information is relevant to the issues in dispute. Questions of relevance are often contentious and loaded with evaluative implications. To allay suspicion that the scientist is clandestinely privileging a particular option, scientists would be well advised to identify as large a range of interpretations and options as possible and to indicate how and why various pieces of science are relevant to each (Weiss 1977). If there is reason for laypersons to waive authority to experts in choosing among these interpretations and options, then those reasons need to be made plain to laypersons. The attempt to arrogate to experts the issues of interpretation and relevance is unlikely to work anyway since some people will insist on thinking for themselves. Moreover, since these issues have normative dimensions, such a move hardly seems reconcilable with democratic political values.

Exploring the common oversimplifications of values talk and facts talk provides a start in mediating agricultural controversy. But by themselves the critical strategies outlined here are not enough. We have explored the shape of stridency; it remains to explore the incomprehension that arises in the conflicting uses of language, and the conflicting sensibilities that underlie and give rise both to strident assertions and to choices of justificatory language.

Notes

1. Schwarz and Thompson point out that problems of this sort are the most vexing in technology assessment and, in particular, make risk assessment problematic (Schwarz and Thompson 1990: Chapter 10).

2. It may be argued that the term "appropriate" as commonly used to represent a certain type of alternative technology frequently begs precisely this question. All too often, the phrase is used to avoid consideration of the very issues of appropriateness.

3. In both tone and substance this early special publication contrasts strikingly with CAST's recent response to the NRC's 1989 report, *Alternative Agriculture* (CAST 1990).

4. In fairness to White-Stevens, the bulk of this literature has appeared since 1972. Issues addressed below are considered, for example, in Lakatos and Musgrave 1970 and Knorr-Cetina and Mulkay 1983.

3

Sources of Incomprehension: The Clash of Contexts

In the previous chapter we took issue with the claims of invulnerability that participants in agricultural controversy made. Some, we found, not only presumed that all others would share their values, but refused to address the substantial problems of going from statements of values to practical agricultural policies. Others refused to recognize the contingency of their facts and refused also to recognize that the relevance of these facts to resolving agricultural controversy depended on evaluative decisions by participants at large. In the chapter our role as authors was a critical one; we wished to challenge, evenhandedly, the assertions of invulnerability made by each side.

But these intransigent claims of invulnerability only reflect deep disagreement; that people are often intransigent (and find in public discourse means of stating and defending their intransigence) tells us nothing about *how* they manage to disagree in the first place. Even if participants fully accepted our criticisms in the previous chapter, it is unlikely that their differences would disappear or even that they would understand any better how they came to differ. At best, they would be able to sustain a civil conversation for a longer time. In this chapter we consider another layer of disagreement, the clash of contexts of argument, or different views of what sort of question is at issue and where one ought to look to find a resolution of it. Such conflicts exist even when there is an absence of invulnerability claims, even, that is, when participants are clearly trying to be open and balanced. They may be most familiar as conflicts between academic disciplines for authority over a particular area: e.g., is agricultural engineering or rural sociology the right court of appeal to assess the impact of new farm machinery? Yet it is probably more accurate to see the conflict as between rival traditions that can exist within many disciplines. These traditions are also something more than expedient rhetorical strategies to promote particular positions; within a single tradition one can find people who will take competing positions, sharing nevertheless a vocabulary, and a body of evidence, concepts, expectations, goals, and rules of arguing that will allow them to engage fruitfully with one another.

While conflict within traditions certainly occurs, we focus here on the more troublesome problem of conflict between traditions, where there is both substantive disagreement and a wide gulf of language and presumption. No longer will our approach be critical; from here on we will be mainly concerned with disagreement based more in incomprehension than in intransigent assertion; our efforts will be correspondingly interpretive and mediative.

Frequently in the competition for authority that characterizes contemporary agricultural disputes each side plays up some ways of talking and belittles others. When the language of dispute is so colonized by each side it is hard to find a means for raising neutral questions, for eliciting dialogue, or for working toward constructive outcomes at all. Since participants chose to discuss issues, it might seem that they must have something in common, for how else could there be anything over which to disagree. Yet with each side so deeply entrenched in its preferred context little common ground is left visible.

To understand controversy in a constructive way we must somehow get beyond the competition of preferred contexts, to draw upon, elicit, nurture, or construct more neutral contexts—contexts in which divergent voices can be brought, without prejudice, into more meaningful conversation with each other. This requirement is a condition of fairness: for genuine mutual understanding all sides must find the playing field level. It is also a precondition of successful mediation and negotiation. Working in this direction requires that we recognize the limitations inherent in particular theories or contexts, separate substance from rhetoric, and seek ways to compare substance without begging key questions. In this chapter we focus on these needs in order to highlight the distinctive contribution an interpretive approach to controversy can make in facilitating productive discussion. Perhaps in the process we can also get some idea of what standards should guide interpretation if it is to serve this aim well.

The Debate over the Family Farm

Perhaps no topic of agricultural controversy illustrates the clash of competing contexts as well as debate on the family farm. Widely recognized as a symbol of national roots, one enshrined in many a farm bill, the family owned and operated farm has for generations now been a shrinking component of U.S. agriculture, shrinking in numbers and in political and economic clout. Many fear it will disappear altogether. As its numbers have declined, however, controversy over its importance has intensified. For some it has seemed the best hope for an agriculture that is ecologically sound and sustainable into the future, socially just, and able to support a healthy and democratic way of life in rural communities (Goldschmidt 1978). For others it is a focus of nostalgia: however desirable the way of living it once provided, the limitations of the labor and resources that most families can now muster have made traditional midsized family farms largely obsolete; they can no longer compete effectively in the exacting economic environment of U.S. agriculture.

There are many ways in which the worth of family farming might be judged: Are family farms economically efficient and financially stable? Do they provide family members a high standard of living or a good quality of life? Are they ecologically more sound or more likely to protect and improve the environment? How technologically innovative are family farmers? Are family farms more sustaining of rural communities? Do they contribute to social justice? or foster political independence? Do they encourage wise management of resources? Do they support valuable rural traditions or uphold important cultural values? Do they represent fundamental human rights or are they essential for human dignity?

All of these standards have been appealed to by one party or another. It might seem possible to resolve debate simply by bringing together all parties in a neutral context where the relative importance of each standard can be decided upon and a compromise effected. But if participants disagree not only about how important these criteria are, but also about how to measure them and even what they mean (and are unable even to talk coherently to one another about what they mean), satisfactory compromise seems out of the question, no matter how good intentions may be. Communication must take place first.

As an example of the way in which competing contexts can undermine good-faith dialogue we consider an interchange that occurred at a 1985 Iowa State University conference on the question "Is There a Moral Obligation to Save the Family Farm?" (Comstock 1987). There Wendell Berry, a Kentucky farmer, teacher, and writer defended family farming against the "industrial values" that are displacing it (on Berry see Part Two below), while Michael Boehlje, a University of Minnesota agricultural economist, offered an economic analysis that questioned whether family farming is better than other ways of farming. Each appealed to a context that he regarded as balanced and neutral, and appropriate for getting a fair answer to a difficult moral question, yet each saw a different context as the proper one, and each reacted sharply to the other's presentation.

Following his earlier treatment in *The Unsettling of America*, Berry presented his argument in terms of the opposition of agrarianism and industrialism. He appealed to a common heritage of cultural values which the family farm represented and which it was the responsibility of all of us to safeguard. So he opened his remarks thus:

> To be asked to defend the family farm is like being asked to defend the Bill of Rights or the Sermon on the Mount or Shakespeare's plays. One is amazed at the necessity for defense, and yet one agrees gladly, knowing that the family farm is both eminently defensible and a part of the definition of one's own humanity (Berry 1987: 347).

Taking a family farm as "one that is properly cared for by its family" from one generation to the next (347), he argued that family farming is good farming in three senses: it is agriculturally good because "there is a ratio between eyes and acres, between farm size and farmhands" that allows "land that is in human use . . . [to] be lovingly used" (349); it is politically good because it supports a democratic distribution of land ownership; and it is culturally good because it "gives work a quality and a dignity that it is dangerous for human work to go without" (350). He thus found it "easy" to show that family farming is good:

> To find evil in it and to argue against it would be extremely difficult. Those who have done most to destroy it have, I think, found no evil in it. They have not argued against it. With them, perhaps, the family farm and family farmer have had the misfortune only of being in the way.
> If a good thing is failing among us, . . . pretty much without professed enemies, then it is necessary to ask why it should fail (351).

This to Berry was the key question and he went out of his way to try to answer it without making pretenses of completeness or stigmatizing any particular group. His answer had to do with our definition of our own humanity, "with who and what we are as a people; the fault lies in our identity, and therefore will be hard for us to see." All of us—consumers and farmers, policy makers, academics and business people alike—were part of the problem. "The family farm is failing," he said, "because the pattern it belongs to is failing. And the principal reason for this failure is the universal adoption, by our people and our leaders alike, of industrial values" (351). The bulk of the essay explored this problem of identity as it appeared among different groups—for example, as "the industrial mind," as the "complex allegiances" of the land-grant system, and as the farmer's "industrial fantasies" (352-7).

What is striking is how much of the work of Berry's argument was done by the context within which he placed the family farm: Shakespeare, the Bill of Rights, etc., elements of our culture that we would never contemplate not preserving. It is also striking how insulated was this context. If we couldn't see how family farms were like Shakespeare, it was because we had acquired an identity that blinded us to the connection. Yet even though the context appeared to be invulnerable, and perhaps incomprehensible, from outside of it (the rhetorical position was a very strong one), Berry was not seeking to polarize debate. He did not accuse those who disagreed with him of being evil, corrupt, or insincere. Nor did he assert the virtues of his position, or attempt to derive it or vindicate it. Context made all that unnecessary.

Yet Boehlje seemed to take Berry's position as quite contentious and definitely biased. Though he did not respond directly to Berry's essay, he made it clear both in his own paper and in his response to Berry's critique of it that he found Berry's context too ambiguous to provide a balanced assessment. In opening his paper Boehlje worried about the reverence shown to family farms, the tendency to "attribute unique, almost mystical, virtues" to them (Boehlje 1987: 361). Responding to Berry's critique, he expressed concern "that the proponents of family farming continue to espouse emotional arguments that are not well substantiated except by selective references to history" (Comstock 1987: 377). These statements doubled as descriptions of Berry's context and criticisms of it as well. That the virtues of family farming were "unique" (if not necessarily "mystical") was precisely what Berry was pointing out. For Berry, measuring with the yardstick of cultural values, it was clear that family farms did have a special significance; for Boehlje, evaluating in terms of various modes of food production, the assertion of unique virtues was highly unlikely, and if the unique values were merely "mystical" they were in no sense praiseworthy, in no sense an argument in favor of the family farm. To the degree that emotional arguments reflected the meaning human beings invested in elements of their environment, Berry *was* utilizing "emotional" arguments. In Berry's context this was nothing to apologize for, in Boehlje's it was.

Boehlje sought a balanced assessment in a quite different context, that of cost-benefit analysis. His declared aim was "modest: to attempt to structure our thoughts as to what the key dimensions of family-based agriculture are, and what economic, social, and moral attributes the family farm actually possesses" (Boehlje 1987: 361). To that end he discussed a wide range of evaluative criteria, including economic efficiency, standard of living, resource conservation, community contributions, and independence. He carefully distinguished family owned farms from family owned and operated farms, and discussed them separately. He presented his findings as tentative—"a set of testable hypotheses rather than definitive conclusions" (363, 366)—and invited proponents of family farms to expand the list of relevant criteria and to "numerically document or logically argue the contribution of family farming to those attributes" (Comstock 1987: 377). It is clear that Boehlje was not out to condemn the family farm (or, for that matter, to laud it). He was less interested in identifying definitive advantages or disadvantages of family farming than in recognizing the interrelations of various elements in family farming and other modes of farming. His remarks were often conditional, sometimes truistic: e.g., "If owner-operators do not charge full cost for their management services . . . while the nonfamily operator does charge the full cost, then costs for managerial services will be higher and efficiency lower with the nonfamily structure" (Boehlje 1987: 366). He cited studies that "refuted" traditional arguments, for example that owner-operators were more likely to adopt conservation practices, but his conclusions were invariably tentative:

> it appears that a combination of tax rules, cost-sharing arrangements, and legal constraints have provided adequate incentives for landlords and investors to adopt conservation practices so that differences between owner-operators and landlord-tenant operated property in the adoption of conservation practices do not exist (368).

Only with respect to independence did he find a clear advantage for family farming, since "it can logically be argued that owner-operators exercise significantly more independence and control over their own future than a tenant-based agriculture" (371).

Boehlje's context was thus the empirical and analytic methods of the social sciences, particularly of economics. With these, he held, we can determine the real relations between factors such as tenancy, family ownership, financial structure, or farm size and the various criteria that have been put forth for assessing the value of family farms.

In the view of Boehlje, and also of Glenn Johnson, another agricultural economist who took part in the conference, this context of social scientific discourse was also the proper one in which to consider the larger issue of the future of family farming. Johnson insisted that the social sciences could contribute to the solution of agricultural problems not only by supplying "value-free positive knowledge" (Johnson 1987: 165) but also by providing normative knowledge and policy prescriptions. Because of the "ethical nature of such involvement . . . social scientists [must] be honestly objective in doing such work." "Ethics," Johnson argued, "requires objectivity."

> Objectivity is particularly vital to the improvement of agricultural policies because so many unobjective assertions circulate on these subjects. . . . both value and supposedly value-free positivistic assertions are made that are not based objectively on experience and logic and, indeed, are maintained despite experience and logic to the contrary. Social scientists bear the special responsibility for testing such questionable assertions and for generating relevant knowledge that has been objectively evaluated on the basis of the tests of experience and logic (166).

While Boehlje's view of the role of social science was narrower than Johnson's, he agreed that objective social science was the neutral context in which to adjudicate family farming issues.

In their subsequent attempts to explain themselves to one another Berry and Boehlje only succeeded in creating a broader gulf between them. Berry chastised Boehlje for allowing "scientific procedure to rule out cultural value as 'almost mystical'" (Comstock 1987: 375). Boehlje replied that "there was nothing in my discussion that suggested that scientific procedure should rule out cultural values. In fact, my whole objective was to propose that we can use scientific procedures to obtain a better understanding of cultural values" (377). Not recognizing how alien were the methods of empirical social science in Berry's context, Boehlje had sought to use them to explore concerns brought forth by all parties. But Berry objected to the very idea of "objective comparison" detached from cultural commitments. The problem was not just in assuming that "cultural values" were empirically measurable, but that Boehlje was "complacently" assuming "that avoidance of commitment assures him the status of a neutral observer" (375). In Berry's view, lack of commitment made Boehlje's work "highly serviceable . . . to the enemies of family farming" (375-6). Yet even then Boehlje did not acknowledge the challenge to his context (indeed, just as Berry did not acknowledge the opposite challenge). If there was a problem with his approach it was in the insufficiency of his data. He refused to countenance (and indeed could not make sense of) objections that came from outside of that context:

> My attempts to be analytical, given very little data or research to draw upon, forced me to develop some arguments based on logic rather than empirical analysis. The appropriate response in such a situation is to

counter the logic and undermine the "speculative results" in that fashion
rather than challenge the results with no analytical framework (377).

Boehlje simply did not see Berry's framework as analytic; it did not have
the characteristics of objective social science.

Part of what Berry meant in objecting to the "speculative" character of
Boehlje's treatment was that Boehlje was dealing only with generalizations about farming; he had not given "a single example of a working
farm." In raising this criticism Berry was identifying a quite different
kind of evidence that he looked for and found compelling, evidence not
for inductive generalizations but in the form of exemplars. To understand the "efficiency of resource utilization and the economic stability of
farm families . . . ," he held, "we need painstaking descriptions of the
workings and the economies of many individual farms." We had to
"answer with responsible particularity . . . all the determining questions
[of what] makes a farm good [which] have to do with structure, proportion, and size or scale—all of which Professor Boehlje entirely ignores"
(Comstock 1987: 376). Boehlje replied that he had not ignored those
issues. But he also did not see "how various descriptions of 'a working
farm' would have significantly contributed to the arguments" (377).[1]

To understand how Berry and Boehlje were managing to disagree, we
need a way of understanding *how* each made sense to himself, a way, that
is, of interpreting the framework in which each was working. Their
contrasting approaches were multifaceted. To begin with, each appeared
to be asking a different question. Boehlje began with a question about
farming in general (i.e., "what kind of farming is good") and explored the
sorts of evidence that would allow a social scientist to say which features
(size, ownership patterns, etc.) were more likely to be associated with
desired outcomes. Berry began with questions about particular farms
(i.e., "what makes a farm a good farm") and explored the kinds of
interactions among humans, nature, and society that made particular
farms the way they were (see also Suppe 1987).

As well as using different questions they used different languages and
different trains of reasoning to address their questions (see Figure 3.1).
Each used a language that integrated empirical and normative questions
in a way that made it difficult to separate them. The central normative
question of what, if anything, was so good about family farming, was not
addressed head on; in the debate its treatment was woven into the
particular framework each speaker was using. Boehlje acknowledged
that his discussion had not answered the question of what specifically
about family farming is desirable, nor whether "these attributes [were] *a
function of* tenancy, family versus corporate locus of control, financial
structure, or farm size" (Boehlje 1987: 373, ital ours). Yet his way of

posing the questions—the search for a "function"—made clear that he saw the attributes that comprise the worth of a family farm (and of family farming in general) as dependent variables, determined by empirically ascertainable factors such as tenancy or farm size, which served as "independent variables." In Boehlje's view, knowing the exact function that related the independent to the dependent variables would both clarify what was desirable about family farming and tell us how farming might be changed to enhance those attributes. The normative and empirical questions were interdependent in the sense that adequate answers to normative questions—what makes farming good, how to make it better—were to emerge from (and were defined in terms of) empirical knowledge, including knowledge of the desires and preferences of actual farmers.

Although he began with the conviction that family farming was good, Berry recognized that not all family farms were equally good. He too could recognize an empirical question about what made them good and he too reasoned in terms of causes determining an effect which could then be evaluated. But his independent variables had to do with elements of culture, the patterns of meanings that guided farm families (or, for that matter, makers of agricultural policy) to make the decisions that led farms to be large or small, owned by families or by absentee landlords, or, in a much broader sense, to be good or less good.[2]

Boehlje's independent variables were thus included in Berry's dependent variables and the central normative question (N) that the conference was to deal with was correspondingly located at a different place for each. Because Boehlje assumed that normative questions could be clari-

Figure 3.1 Competing Conceptions of the Problem of the Family Farm

	Assumptions	Independent Variable	Dependent Variable	Problem to be Solved
Boehlje	social science good	empirical attributes (e.g. farm structure)	quality of farm system (N)	what kind of farming is best?
Berry	family farm good	cultural attributes (e.g., "industrial mind" or not)	empirical attributes of farm and other satisfactions of farm living	what makes a farm a good farm? (N)

fied by empirical measurements, he needed to define the desired ends of family farming in terms precise enough to allow comparisons of various kinds of farming. Thus, his assessment was keyed to terms such as "standard of living" with well developed technical usages in social science. Because Berry assumed a normative definition of family farming that implied its goodness, his concern was to illustrate what made it good rather than to compare it with other kinds of farming. For empirical support he cited examples in which a strong and stable culture produced responsible traditions of farming that made life meaningful and worthwhile for farmers. Although he recognized important relationships between empirical attributes such as farm size and the goodness of family farming (e.g., farms must be small enough to be "lovingly used"), little could be accomplished by trying to analyze that goodness (or "quality of farm life" or the like) in terms of a few empirical factors. Because they reflected the quality of an individual's life in a particular culture in such a complicated way, varying those factors in the absence of intimate knowledge of a particular farm could not guarantee a high quality of life.

Such differences in terminology and reasoning were far more than stylistic. They had far-reaching implications. For Boehlje "income generation capacity" (Boehlje 1987: 367) of farms was likely to be a crucial measure of farming success, while Berry suggested that even in a farm "that does not entirely support its family . . . , the values of the family-owned and family-worked small farm are still available both to the family and to the nation" (Berry 1987: 348). The different patterns of reasoning also led to different stances toward available evidence. Boehlje, for example, cited evidence that owner-operators are not "more likely to adopt conservation practices" than other farmers (Boehlje 1987: 368), while Berry implied that family farmers had been operating under such adverse conditions that often they could not take very good care of their land. From his point of view studying the behavior of family farmers under the current circumstances would have been very much like doing chemistry with impure substances; were we to study family farmers in a proper cultural and economic setting, we would find them taking better care of their land.

This interchange reflects well the complementary blind spots that can arise when people address an issue from competing but very different contexts or frameworks. Neither Berry nor Boehlje saw his context as contentious. Berry simply did not see any point in analyzing statistically the complex relations which, in the view of social scientists, determined various characteristics and outcomes of family farming. The value of the family farm was something like the value of the works of Shakespeare; discovering the mean income ten years later of persons who read *Hamlet* when they were high school seniors told one nothing about how that reading helped define their humanity, nor did it have anything to do with the reasons Shakespeare's plays were preserved, read, and performed. In short it told one nothing about the value of the works of Shakespeare. Boehlje did not see cost-benefit analysis as contentious; on the contrary it was the essence of rational appraisal where the planning of social policy was at stake. Such analysis did not ignore achievements on individual farms; their true weight and nature were made apparent by the process of social scientific analysis. In future, as techniques of measurement improved, we might become aware of more subtle elements of family farming that made it especially valuable. But talk of "cultural value" or of the "industrial mind" displaced from the land seemed only eloquent ignorance. It obscured what needed to be understood and could be permitted no force in a rigorous analysis. Only objective social science could ultimately lay the issues to rest.

What Interpretation Can Do

Misunderstandings like these are not only sources of polarization, they are also impediments to understanding deeper differences in controversy. When they fail to recognize or acknowledge the thrust of opponents' objections and instead reassert their preferred contexts, disputants isolate themselves from engagement with the deeper issues. Berry and Boehlje were not able to find a mutually acceptable way of formulating their differences. The central issues that divided them—of what sort of knowledge was needed, of how it would be had, and of exactly what problem it would be used to solve—were relegated to the margins of the debate and the issues were never joined very fully.

Yet deeper issues can be joined, if not always resolved, through the mediative role of balanced interpretation. Interpretations can often be fashioned which bring the substance of opposing positions into contact without begging key questions or offending sensibilities (Shepard 1989). A crucial step in the process is to identify the contexts in which disputants are operating, as we have done above. Highlighting Boehlje's appeal to social science and Berry's appeal to agrarian exemplars was a way to illuminate the discontinuities that ran through their interchange—the ways they managed to talk past one another. Focusing on these discontinuities, drawing out matters that are implicit and linking them to points of disagreement, allows participants to talk *to* each other instead; interpretation gives disputants the chance to try to understand their differences rather than react to each other from half-conscious feelings. And as well as aiding in the management of controversy, interpretation can also help participants grapple more successfully with the intricacies of the questions they face. Some of these are the questions we raised critically in the preceding chapter—normative questions of what standards or criteria are right or appropriate, and why, and how they will be applied, and empirical questions of appropriate inference, which must include the hazards of mistaken inference. But they will also include questions that are indivisibly both normative and empirical, questions that are sometimes hidden in the sort of language disputants use to try to clarify the issues.

But if a probing interpretation can open the way to discussion of deeper differences, it can also backfire: if it begs questions dear to the heart of some disputants or otherwise fans the flames of acrimony, interpretation can heighten polarization. What standards then must guide interpretation if we are to prevent such outcomes?

Beyond Fixed Standpoints

In attempting to move discussion onto constructive paths most people would agree that interpretation of the views of participants in controversy needs to be fair and impartial. Yet different people attach different meanings to those terms. What is a fair assessment to one may be outrageously partisan to another. Boehlje and Berry each went out of his way to present a fair assessment of the family farm, but each balanced his assessment in terms of a different center of gravity. Each sought a fixed reference point outside the immediate context of the family farm debate from which to assess claims made about family farming. Berry began with what he regarded (probably rightly) as the widely held conviction that family farms were good, all things being equal. He took up the problem of what it was about human beings and the institutions they created that had undermined the realization of this good. One of these undermining influences was the belief that "avoidance of commitment" was the proper stance from which to assess family farming, precisely the framework within which Boehlje was working. For Boehlje the fixed point was the objective methodology of social science, so his main objection was that Berry espoused "emotional arguments that are not well substantiated." In short, Berry objected to Boehlje's epistemological standpoint; Boehlje objected to Berry's moral standpoint.

In such a situation what's an interpreter to do? The idea of providing a third fixed point, that of a detached and neutral, yet skilled interpreter who can recognize what's really going on, is unlikely to get us anywhere. Because the two views are grounded in such disparate ideas of a reasonable response to the issues, the interpreter will be much taxed simply to describe what's going on. The gulf in languages and logic separating the participants is so wide that it's not clear where we are to find a satisfactorily neutral language or an ultimately fair way of rendering arguments. Talking on the one hand of the family farm as something like the works of Shakespeare may well be unfair to Boehlje, for the approach he takes might simply deny the validity of the analogy. Talking on the other hand of culture as something like a "dependent variable" may well be unfair to Berry; he might simply reject the use of the term in this case.

At stake here is the nature of impartiality itself. Once two such "fixed points" have been introduced into the debate, adding yet a third fixed point promises only to compound the problems. Genuine impartiality would seem to require that one's interpretations avoid either assuming or presupposing any claim that is contested in the controversy, whether it is directly disputed or implicated in some less overt way. To violate this rule would be to invite the charge that one has taken sides in the dispute; one of the parties will have reason to feel it has been unfairly treated. And if the aim of interpretation is to engage participants in more constructive discussion, such bruised feelings are anything but helpful.

The range of perspectives in contemporary agricultural controversy means that these kinds of problems will be unavoidable. They highlight the need for standards of interpretation that are practical rather than theoretical and relative to the context of controversy and to the actual views and understandings of participants.[3] In this sense of impartiality the proof is in the pudding: Has interpretation brought the disputants to more meaningful and constructive conversation? *Their* answers to this question are crucial. It is the reception of interpretations that determines whether they are helpful, not the recipe or intent behind them or how they measure up to some external yardstick. What this means is that the interpreter becomes a conductor of conversations rather than a teller of truths; and the interpretations become leads and cues to synchronize point and counterpoint more meaningfully and constructively.

Impartiality Relativized

What can make such contextually grounded interpretations succeed? Interpreters will need some guidelines to use in conducting conversations, some notion of what a helpful mediation will be, and some reason to think that it will also be perceived as fair. Four practical concerns have guided our interpretations of agricultural controversy:

1. *Have we treated each side equally, applying our approaches to interpretation symmetrically and evenhandedly?* In whatever ways we probe or unearth the subtler meanings of one side, we must give the same scrutiny to the other side as well. "Turnabout is fair play."

2. *When confronted with our interpretations of their views, will disputants feel understood?* Will they be able to recognize their views in our interpretations, and perhaps even find new insights? Will they be able to own our interpretations publicly? In trying to move discussion forward, the burden is on the interpreter to be faithful to the actual beliefs and sensibilities of participants. But this cannot mean leaving interpretations in the partisan language of participants. To facilitate dialogue, interpretations must be intelligible to those who do not hold the view interpreted as well as acceptable to those who do. In this sense the interpreter of controversy is also a translator and must balance needs for fidelity against sometimes conflicting needs for intelligibility. If translations are too slavish in following the idiom of proponents, they may often appear to opponents as having perpetuated the obfuscations they would challenge. But if the translations are not very careful or reach too far toward the partisan idiom of opponents, then proponents will likely reject them as distortions. It seems unlikely that initial interpretations (such as those offered later in this book) will achieve the desired balance, but there is reason to hope that an iteration of interpretations and responses to them will effect translation, especially when there is an evenhanded effort to make each view transparent within the idiom of the other.

3. *Have our interpretations imposed meanings on the dispute that come from outside of the context of the controversy?* Have we avoided the temptation—one so endemic that we've come to think of it as the interpreter's bane—to appeal to ideas, meanings, standards, or assumptions that are not indigenous to the dispute—to external "fixed points" (e.g., a Marxist framework, or an idealist framework) that will not be accepted by all the parties? There is a burden on the interpreter to be faithful to the context of controversy. This does not mean that things can never be imported from outside, but it does mean that what is imported must pass a stringent test of relevance: Can the new ideas be shown relevant to the dispute with arguments that are intelligible and acceptable to all the parties? In practice what this test does is to restrict the interpreter to working with the dispute essentially on its own terms using the understandings and verbal know-how that are already represented in it.

4. *Will the participants feel they have been treated fairly?* This question applies to the whole treatment and implies some sensitivity to participants' own standards of fairness. It is not enough that

interpretations be evenhanded and reasonably acceptable piece-meal. They must also reflect the coherence and integrity of each view and the distribution and weighting of its concerns, and they need to do this in ways that are recognizable and acceptable to those who hold the views. If any of these conditions are not met, participants will not feel they have been fairly treated.

Such may be the characteristics of an effective mediation, but is it also an impartial mediation? The problem in answering this question is that almost inevitably participants in controversy will have different ideas of what impartiality means. Some may look to ethical principles, such as the principle of utilitarianism—to do what serves the greatest good of the greatest number—or Kant's categorical imperative—to do only those things that you can consistently will that all people should do. Others, appealing to the legacy of logical empiricism, may conceive impartiality as that property possessed by scientific knowledge claims of being independent of moral or political value judgments. But here again, because there is so great a diversity in conceptions of impartiality, we must use a quite different conception, one more modest and more pragmatic (Figure 3.2).

The strategies to achieve impartiality we outlined above are rooted in common sense rather than transcendent principles and draw their meaning from the context of controversy and the actual beliefs and sensibilities of those who participate in it. Impartial interpretation, in this sense, is a situated achievement. It must be reworked in each new set of circumstances. As such it is quite distinct from ethical commitment, for it must be refabricated to meet the unique mixes of ethical commitments and moral standards that arise in each new controversy. But unlike the logical empiricist's value neutrality, impartial interpretation is not a strictly

Figure 3.2 Senses of "Impartial"

Sense	Source	Test
Ethically Right Action	ethical commitment	agreement with principles
Value Neutral Knowledge	scientific method	analysis of presuppositions
Fair Mediation	common sense, context	participant perceptions

analytic matter either. To check whether a knowledge claim is value neutral one analyzes its warrant to see whether it presupposes or implies value judgments. But to check whether interpretations are impartial, one must see how they are received by participants in controversy. The closest we can get to an analytic guide for the interpreter is the above rule not to beg contested questions—including questions of what is impartial interpretation. Yet to use this rule one has to make difficult practical judgments in figuring out just what is contested.

In seeking ways to enhance open discussion of agricultural policy we wish to augment the common approaches of answering objectively and advocating effectively with the distinctive contributions that come from interpreting impartially. It is this approach which we find most needed and much neglected today. We do not see this as a way of eliminating disputes or the subtler tensions that arise from the conflict of disparate contexts of language and reason, but we believe that it can help people learn to work with those tensions constructively to reduce acrimony and facilitate dialogue.

This chapter has brought us closer to recognizing the depth of disagreement in agricultural controversy and it has indicated how interpretation might be used to plumb those depths. But such tools of understanding are limited; we can recognize that Boehlje and Berry did appeal to vastly different contexts, but we still know little of what made those contexts so compelling for them, of how their participation in the family farm controversy was integrated into the way they perceived and interacted with the world and understood their own lives and situations. For that we need a more powerful concept than "context," a concept of "ideology," developed in the next two chapters.

Notes

1. Similar arguments about the comparative validity of agricultural science by generalization or by exemplar have arisen in response to the National Research Council's report, *Alternative Agriculture* (NRC 1989), which utilized case studies. They arise in general with regard to farming systems research (Batie 1990; Busch 1989b; CAST 1990; Hileman 1990a).

2. Berry of course did not use this language. We use it here quite loosely, without its usual presumption that "variables" are empirically measurable.

3. See Shepard and Hamlin 1987, and Shepard 1989. It is in this regard that we differ most clearly with the recent work of Schwarz and Thompson (1990); compare their Chapter 1 and pp. 106-109 with what follows here.

4

From Values to Ideologies

In the last chapter we saw how people seeking to confront an issue openly and to engage with opponents fairly and honestly still managed to talk past one another. Not only did they disagree, but each was also apparently unable to comprehend how the other was approaching the issue so differently, how they had managed in particular to choose such a strikingly different context from which to proceed. This led us to consider the problem of how one might try to bridge such differences to render each view more intelligible to the other. Since each side tends to want to defend the language in which its proposals make sense, an interpreter is hard pressed to find neutral translations; one can only hope for translations, it would seem, that are minimally yet equally disagreeable to both sides.

To leave the problem here would be unsatisfactory. The concept of mediatory impartiality does give one a sort of formal guideline, and it does enjoin the interpreter to respect the content of disagreement, but it provides no idea of *how* or *how far* to respect the substance that is to be translated. While the interpreter might be able to split the differences of language and context, she would not know what sacrifices she was requesting, in inviting participants to accept a particular translation. She would not know (nor would participants necessarily know) what it was about each side's language that made it so compellingly attractive. Nor would the interpreter find in the notion of evenhandedness any indication of what kind of interpretation would be helpful, what sort of evenhandedness would be perceived as such. Which elements in the rhetoric of participants might they comfortably abandon? Which will they cling to stubbornly, come what may? The rhetoric people use in controversy is often something more than just a way of talking. So one must ask what more it is; how is it that people manage to select the same sorts of language, context, and rhetorical strategy over and over. How, for example, are the similes (family farms are like Shakespeare's plays) or maxims (the methods of social science are the best way to inform normative choices) that exemplify rhetorical choices integrated into coherent ways of interacting with the world.

In this chapter we ask what manner of coherence lies behind the rhetoric that participants use in controversy and we answer that it is the coherence of "ideology." In contemporary discussions of matters of science, technology, and public policy, "ideology" is a widely used term, though one often used in vague and conflicting ways. Yet the term remains useful: some of these conceptions of ideology may help to make sense of the range of differences that are implicated when parties to dispute talk past each other. With such guidance perhaps one can begin to understand what gives shape to the conceptual discontinuities and rhetorical gaps that are endemic to agricultural controversy and other areas of policymaking as well.

Competing Rationalities

Besides involving accusations of bad values or mistaken facts, or reflecting mutual incomprehension, agricultural controversy also involves accusations of irrationality. Some critics of the agricultural research establishment, for example, have been dismissed as wilfully irrational, forsaking accepted standards of judgment in order to push for their pet policies. Johnson and Wittwer, for example, remark that "as advocates, activists often sacrifice objectivity to promote prescriptions that they put beyond investigation and research" (1984: 9-10). While Johnson and Wittwer see a legitimate place for criticism and advocacy, their remark reflects a widespread tendency to find something irrational in those with whom we strongly disagree.

It seems unlikely that many people choose to be irrational when they participate in discussions of agricultural policy. While their presumed irrationality may reflect misinformation or misunderstanding, or a conscious strategy to call attention to one's views in a political forum, it might also mean that they are operating with different standards about what is realistic, natural, rational, objective, proper, moral, or sensible. Perhaps then, what we are seeing are competing worldviews or competing rationalities (Schwarz and Thompson 1990: 11-13, 59ff; Cotgrove 1982: 34; Beus and Dunlap 1990).

Another example of a deep disagreement in agriculture will illustrate what we mean by talking of a clash of rationalities. C. Dean Freudenberger, an advocate of "sustainable agriculture," makes a distinctive appeal:

The history of civilization's attempt to feed itself is dismal. It is one of massive and irrevocable loss. . . . The statistical data about contemporary U.S. agriculture is simply an additional though alarming chapter in this historical record.

The haunting question before us is this. Is there any possibility, particularly during our moment of human history, where human populations have expanded and continue to expand beyond all boundaries of imagination and where our agricultural technology has become so powerful and momentarily productive, to reverse the historical record? . . .

The thesis of my work is that we can come up with an answer to the question. It lies in the idea of sustainability in agriculture (1982: 622).[1]

A contrast to Freudenberger's perspective is the view of R. C. Rautenstraus in a paper on "Public Responsibility of an Agronomist—A University President's View" (1980). Rautenstraus congratulates agricultural scientists on carrying out "the most successful scientific inquiry" ever made, which has made the most "profound, sweeping, and lasting changes . . . [and] so thoroughly reshaped the world that it is impossible to conceive of an ordering of modern civilization that is not underpinned by its contributions." In Rautenstraus' view, "the Green Revolution . . . has taken hold and the signs are now unmistakable that a cycle of poverty, disease, and ignorance that is as old as time is coming to an end." In the U.S., scientific agriculture has "freed the country from the threat of famine and insured the population of an ample diet . . . without requiring that a disproportionately large share of the work force remain in agriculture."

It is safe to say that these people disagree. But how? While both statements make up part of an argument, neither is structured as a formal argument. Both are descriptions of the existing situation which set the scene for the argument each author makes. Neither reveals much about the internal train of thought that drives each author to his conclusion, neither seems to depend on any master metaphor or axiom. The authors may hold different values, but that is not clear. Yet apparently they look out at the world and see it very differently. We might be tempted to say that one sees the cup half empty while for the other it is half full, or that Freudenberger is a pessimist (in fact he is not) while Rautenstraus is an optimist. But these characterizations don't get us very far. Rather than trying to reduce their disagreement to its particular genus and species it seems more useful to enlarge on their statements to see if we can characterize the moving picture of history that each is relying on. Each is concerned with representing our situation in the historical present; each sees history as a vector describing where we are going at present. But their vectors point in opposite directions (see Figure 4.1).

For Freudenberger, history reveals impending calamity. In our lack of concern for the sustainability of agriculture (and for the survival of humanity) we are like a bathing party on a summer afternoon rafting down the Niagara River a few hundred yards above the falls. The imminence of our plunge, obvious to anyone who stands up to look, is reason enough to take immediate action. But the afternoon is warm and pleasant, and most of us are too languid to rise; after all, people like Freudenberger have said this sort of thing before. Nevertheless, the choice we have is to stir ourselves to act now or to float on to oblivion in our current stupor.

For Rautenstraus, by contrast, history reveals the approach of an age of sufficiency, prosperity, justice, and freedom. It is as if we are building a stairway from a hostile desert up the side of a cliff to a fertile plateau. We have nearly reached the top and have already begun to reap the bounty it promises. Scientists are our master builders, their designs have guided the project and they are showing the rest of us what we must do now to continue. Now and then some of us become disillusioned and urge abandonment of the project; perhaps we should do the best we can in the arid and rocky terrain at the foot of the cliff. But we have to get to the top! There are too many of us for a just and prosperous society to survive in the lowlands. Our problem is one of maintaining faith, for unless we have the faith to keep supporting the scientists we never will reach the top.

What the statements of Freudenberger and Rautenstraus have in common with these vignettes is a focus on the meaning of the present—a concern with what is significant in how we have come to this moment, where we need to go in future, and what should occupy us now. Because both authors are assessing significance, the question who is more accurate is beside the point—both necessarily select facts, including some, leaving out others. Freudenberger focuses on soil loss but leaves out widespread

Figure 4.1   Competing Images of the Present Historical Situation

	Freudenberger	*Rautenstraus*
Impending Prospect	calamity	prosperity
Appropriate Response	radical action	perseverance
Main Obstacle	complacency	lack of faith
Appropriate Agent	any who have seen the threat	scientists

malnutrition. Rautenstraus highlights increased food production but omits soil erosion. One could say that differences in their values determine which facts each author includes and excludes, but far from answering the question that would only make it less accessible, for the problem is precisely to understand how meanings lead authors to value some facts more than others.

One can also see their divergence in their rhetoric. The words and phrases they use create meaningful pictures, bypassing the need for explicit argument. Freudenberger labels modern agriculture "momentarily productive." Most would probably agree that it is currently highly productive and that it is only recently—a moment ago in the human past—that it has become so. But "momentarily" goes further to suggest that the moment will soon end and we will go back to what we had before or worse. Rautenstraus notes that a high level of agricultural production has been achieved "without requiring a disproportionately large share of the work force to remain in agriculture." No one denies that the proportion of American workers in agriculture is lower than it was, but it is not clear how to distinguish a "proportionate" share of workers in agriculture from a "disproportionate" one. In both instances the language ushers us around questions that might challenge the author's readings of significance. Each evades scrutiny of his picture by pointing us away from the picturing to look at what is pictured.

Are either of these authors being irrational? Do they wilfully sacrifice objectivity, for instance? Even though their selectiveness in choosing facts and using language is fairly transparent, we do not think such charges are appropriate. Even were they concerned only with describing the course of agricultural history, they would need to select what was significant from a vast array of events, and they would need some standards of significance to guide that process. But they seek further, looking to history for the meaning of the present and for the course of future policy. They seek warrant for a course of action with which to confront the unknown and so they must find some kind of framework with which to make clear the appropriateness and relevance of policies. Perhaps Freudenberger's and Rautenstraus' alternative "histories" are not irrational, but rather are somehow necessary to cope rationally with the uncertainty of the future. Perhaps what their differences reflect are different frameworks for projecting from the past into the future.

The Liabilities of Values Talk

Why cannot these differences of framework be regarded simply as reflections of different sets of values? This move is tempting, and certainly would be simpler than getting lost in the vagaries of "ideology" or similar terms. Freudenberger, one might say, places a higher premium on the future value of soil resources than does Rautenstraus, while Rautenstraus places greater importance on current food sufficiency.

Such interpretations, however, can be misleading. Rautenstraus might argue that he is not discounting conservation, that Freudenberger is just wrong about the causes and implications of soil erosion. Or Freudenberger might argue that he is not downplaying the value of having enough food now, but that technologies that exploit resources to increase production don't necessarily contribute to getting the food where it's most needed. In this way it remains unclear whether value differences can be isolated from the hodgepodge of intersecting judgments and concerns that make up disagreement, or, if they can be, whether they actually account for the disagreement. Freudenberger and Rautenstraus may well hold different standards for assessing agricultural technologies and their priorities for development in agriculture are certainly different. Yet it could be argued that each starts from the same basic value of the survival of civilization, though they differ about what is needed to realize that end.

Nor is it clear that pursuing such a train of analysis would pay off in the end, for "values" talk both prompts and masks a welter of ambiguity and confusion (Kluckhohn 1951; von Wright 1963; Winner 1986; Shepard 1988). Either it cuts too finely by focusing on personal idiosyncrasy —values as personal attributes—or it cuts too roughly by presuming a greater commonality than is warranted—we may all value "peace" and "justice," but judging from recent controversies there is little agreement about what these mean or how to bring them about. "Values" talk either overemphasizes rational self-direction by focusing only on deliberate thought and action (e.g., "farmers will continue to use pesticides because they value present income over the quality of the environment") or it overemphasizes unreflective, habitual assessments by focusing only on the tie between subjective feelings and expressed preference (e.g., "65 percent of the farmers sampled expressed a preference for working land they owned").[2] Moreover, to keep the term "values" in the central position would be continually to risk the oversimplification and the privileging of one's own assumptions that accompanies the popular distinction between facts and values.

The sort of coherence that we find displayed in contemporary agricultural controversy is more general than personal attitudes but less general than whole cultures (Dumont 1980). The integrity of the outlooks of Freudenberger and Rautenstraus (and of many others who will identify with the positions they take) is reflected in much more than differing values—it appears in the ways they see the world, their reasons for engaging in policy dispute, their visions of a desirable future, and even their definition of their own place in society, history, and nature. Each of these—worldviews, rationales, visions, and self-definitions—involves values, yet each is clearly something more or other than a system of values. Each reflects determinate forms of rationality in ways that emotional outbursts do not, yet none is free from ties to subjective feeling. While each treats a broad range of topics, its concern with them is often limited by partisan purposes, and many topics of assessment may never arise. Moreover the frameworks themselves are in part products of entrenched controversy: because they may evolve in unexpected ways during controversy, such frameworks are not reducible to stock listings in philosophical or social scientific catalogues of alternatives.

Why "Ideology"?

For these reasons "values" is inadequate. We can use a variety of more precise terms: "value judgment," "standard of evaluation," "prescription," and the like. But our focal term must be "ideology," a term both more specific to our purposes and more apt in serving them. Although "ideology" may seem quite as vague and ambiguous as "values," its vagueness and ambiguity fit our problem better: "ideology" belongs with "politics" and "controversy."

The origins of the term illustrate the sorts of tensions that have shaped its associations. Introduced by Destutt de Tracy in the aftermath of the French Revolution, "*ideologie*" was the name for his science of ideas, which aimed to reduce all ideas to their origin in sensation and thereby to clear the mind of the speculations that had bolstered the old regime. De Tracy sought to put science in the service of political enlightenment. With fellow "*ideologues*" he devised a system of education to transform France into a "rational" and "scientific" society. To Napoleon, however, the iconoclasm of the ideologues was dangerous. He linked the terms "*ideologie*" and "*ideologue*" to the worst features of revolutionary thought, and succeeded thereby in turning them into terms of disapprobation. The current English term "ideologue," along with its cognates in most other European languages, retains this stigma (*Encyclopaedia Britannica* 1984, 9:194; Mullins 1972: 499 fn 6).

But the term "ideology" remains ambiguous. In the works of Hegel and Marx and of Weber and Mannheim, in the tradition of Anglo-

American sociology and in European "critical theory" (to mention only a few sources) one finds a daunting array of conceptions of ideology. Yet there do appear to be two broad, usually opposing purposes to which treatments of the term have been shaped.

On the one hand are those who would use the term for purposes of explanation: to explain social arrangements and political behavior scientifically. In the work of Weber, of Mannheim on the sociology of knowledge, and in much empirical sociology and political science, ideologies are the socially situated beliefs that shape or legitimate action and programs for action. Some (Weber) find explanatory power in the beliefs themselves; others (Mannheim) vest explanatory power primarily in the social bases from which ideologies come and which are often held to determine them. Thus we might speak of an idealist and a more materialist or structuralist wing among explanatory theories of ideology.

On the other hand are those who would use the term as a central concept in social criticism, like Marx and the critical theorists, whose ultimate concern is with the transformation of social and political systems. In these treatments ideology is understood as "false consciousness." It is a systematic inversion of the actual historical situation that distorts and deceives in the interest of a dominant group. Implied, of course, is the critic's own access to the high ground of correct appraisal.

There is much in these broad and sometimes intersecting streams of thought that is worth borrowing for our purposes. But no one treatment quite fits our needs. Perhaps that is not surprising, for our concern is neither to explain nor to critique, but rather to interpret, explore, and mediate—aims largely ignored in the evolution of the concept. Indeed, so thorough has been the neglect that it would be impractical to try to build here a systematic definition of "ideology" to meet our needs. One can, however, extract from the works of various theorists the rough features of a mediatory conception of ideology—a definition that puts us in "the right ballpark" as it were. Writing, for example, "On the Concept of Ideology in Political Science" in 1972, Willard Mullins proposed to define an ideology

> as a logically coherent system of symbols which, within a more or less sophisticated conception of history, links the cognitive and evaluative perception of one's social condition—especially its prospects for the future—to a program of collective action for the maintenance, alteration or transformation of society (510).

Mullins' definition recognizes five crucial features of an ideology: (1) historical consciousness, (2) cognitive power, (3) logical coherence, (4) evaluative power, and (5) action orientation (507). Each of these helps demarcate ideologies from other cultural phenomena, so we consider them more closely.

In the contrasting descriptions of Freudenberger and Rautenstraus we have seen already something of the manner in which historical consciousness is involved in agricultural controversy. In their opposing stories of where agriculture has come to and where it might be going, each recognized a strikingly different historical present. Each sought to project the changes he perceived into the future and to take charge of the outcome. Thus with Mullins we could say that in their "historical consciousness . . . the shape of the future, the nature of historical change, and the limits and the possibilities of human control over these changes, become questions of overwhelming importance" (504).[3]

The cognitive power and logical coherence of ideologies are closely related. They were less visible in the passages from Freudenberger and Rautenstraus since we focused only on short passages from the works of these authors. Cognitive power refers to the way in which ideologies channel thinking. The anthropologist Clifford Geertz sees ideologies as meeting our real needs for explanation and understanding, but with a practical slant. They "render otherwise incomprehensible social situations meaningful, [allowing us] to so construe them as to make it possible to act purposefully within them" (Geertz 1964: 94). According to Geertz, ideologies are especially important in times of large social changes (such as the restructuring of American agriculture) when traditions (such as family farming) are challenged or displaced. People recognize a need to act, and they seek a framework to guide their actions, to help them figure out what to do. The more leisurely and systematic approaches of social science do not meet such needs, for the needs are too urgently practical, too immediate. We cannot meet them without recourse to the selections and simplifications that ideologies provide, but that science eschews.

Patterns of selection and simplification give an ideology its cognitive power. But cognitive power need not come at the cost of logical incoherence or inconsistency. Indeed, Mullins sees the "constraint" that ideologies provide as distinguishing them from "random or inchoate ideas, attitudes, or feelings about politics" (507, 510; see also Converse 1964). This constraint is not necessarily formal. The logic of ideologies, Mullins notes,

is usually as broad as the conceptions, reasons, and justifications that typically "count" in the sociocultural structure . . . within which the ideology operates. Nevertheless, given these variegated logical resources . . . , ideology must not repeatedly violate their canons of sensibility. Within their confines the ideology must "make sense" and not result in logical absurdities (Mullins 1972: 510).

Ideology is thus distinct both from "the ad hoc, piecemeal appeals of propaganda" and from inarticulate "ethos" or "spirit of the times" (510).

The practical import of ideologies is reflected most directly in the last two of Mullins' characteristics, the evaluative power and action orientation of ideologies. In "making sense" of a baffling world, ideologies also evaluate "the contours of reality," as it exists, and also as it "might be shaped depending on the intervention of politically organized human beings in the historical process" (508). We saw such evaluations in the descriptions of Freudenberger and Rautenstraus—one saw the present as threatening, the other as promising; one saw need for fundamental change, while for the other it was perseverance with present projects and commitments that mattered most. For each author an ideology was helping to explain the world, to discern what mattered, to judge it, and thus to call one to action. The ideology did not determine action, but it did help to provide a warrant for action. As Mullins put it, the significance of ideology was not that it "causes one to do" but that it "gives one cause for doing." With purposes and evaluations, actions become meaningful; one has reason to act and can make sense of action both to oneself and to others (509).

Mullins' definition indicates what kind of thing an ideology is, but by itself it does little to help distinguish or identify particular ideologies. For the interpreter or mediator, however, these are the more pressing needs. To address them one must look beyond the formal definition of "an ideology" and suggest, as Mullins definition does suggest, the sorts of features that might be useful in describing or distinguishing particular ideologies. Here is a concise list:

1. *Worldview*—How is the world perceived? What goes on in it? and what does it mean? What descriptions and accounts are accepted or asserted?

2. *Ontology*—What is real? What about the world doesn't budge?

3. *Epistemology*—What can be known? What does it mean to know it? How can it be known? What tests of knowledge matter?

4. *Normative Outlook*—What needs to be evaluated? and according to what standards? What priorities are held? What tradeoffs are made? What scruples or constraints are observed? (Aiken 1985)

5. *Agency*—What sorts of things can act in the world? Where, when, and how does action take place? What kinds of action are effective or responsible?

6. *Vision*—What future is desirable? What obstacles threaten to keep us from it?

7. *Program*—What policies are recommended? What is the agenda for action?

8. *Political Theory*—How are practices to be justified? How should power be exercised or challenged? When and in what forms is social authority to be trusted?

9. *Self-definition*—How are we to understand our place in the world? What are we like? What could (should) we be like?

10. *Social Structure*—Who are the significant others with whom we must interact? What are they like? What will they do?

Questions of these sorts have guided the descriptions of ideologies that appear in later chapters. Although they could be subdivided more extensively and rearranged or rephrased, the rough conception of ideology they embody would not change much. So long as it takes in key elements of what the world is like, who lives in it, what's good, bad, changeable, who we are, what we are to be doing, and how we can know and act, a list of questions will hit the main features of ideology.

Liabilities of Ideology Talk

Just as values talk brings with it a host of problems, so does ideology talk. The problems are especially vexing to anyone who would put "ideology" in service of impartial interpretation and mediation. There is great danger of sliding into a pejorative use—a twofold problem: one's audience may have difficulty understanding how the term might be used nonpejoratively and one must take pains to ensure that one's own descriptions of ideologies do not turn into critiques.

When people hear their views labeled "ideology" they would be well "advised to reach for their pistols," suggests the social philosopher Charles Taylor. "The chances are, no good is intended" (1983: 39). Geertz quotes a paradigmatic parody: "I have a social philosophy; you have political opinions; he has an ideology" (1964: 47). The parody is "familiar" he says, yet the attractions of a term with which one can quickly stigmatize an opponent's views are hard to resist.

Examples abound. For one in agriculture we might take Tony Smith's attack on the neutrality of social scientists at the 1985 Iowa State conference on the family farm. Characterizing "ideology" as "a technical term used to describe assertions that claim to be objective, impartial, and scientific, but that actually serve the interests of ruling groups," Smith charged "academic experts connected with agriculture" with shifting "back and forth from relatively straightforward empirical assertions to more or less crass apologetics for the corporate system of agriculture" (Smith 1987: 176). While creating an "aura of scientific objectivity," this "style of speech" had "often masked a style of thinking where ideological considerations" fused indiscernibly with scientific ones. Smith identified four "discourse mechanisms" that generated such effects: "the exclusion of relevant questions, the omission of known facts, the retreat to abstract models, and the failure to consider relevant power relations" (176-7). To Glenn Johnson's charge that Smith had used the same "methods of discourse" in an equally "biased manner," (Johnson 1987: 188) Smith asserted that it was precisely his point that "inherently political issues *cannot* be successfully depoliticized, . . . my own presentation . . . [does have] a political viewpoint." For Smith the difference was that his "political viewpoint" was explicit while the opposition's was cloaked in "ideology," and pretended to be social science (Comstock 1987: 196).

Since Smith spoke as a critic of the agricultural establishment, his use of "ideology" was aimed precisely at exposing to social scientists the political implications of their own enterprises. His pejorative use of the term was also performative; like a slap in the face it insulted and challenged. Small wonder Johnson found it polarizing. For precisely the reasons Smith found "ideology" a useful term, a mediator will find it a difficult one to use.

It may even be asked whether it is possible to use the term impartially, not because listeners will invariably hear the term as an accusation, but because much of the business of description and translation (identifying simplifications, for example), can so easily be taken as accusatory or critical. Take Smith's argument again. At the heart of his distinction between political views and ideology are issues of explicitness and straightforwardness: when one's views are owned openly they are free from the stigma of "ideology," but when they are hidden or covert, they are not. Such political conduct is improper or unfair, Smith seems to hold. It may mislead or deceive, and thus undermine necessary and healthy political discourse. But to apply this argument Smith must have clear ideas of the views different people really hold, of what the real political issues are, of what is open and what is hidden. He is assuming, that is, some way of marking where acceptable politics leaves off and ideology begins. Yet such an assumption presumes on other people; Johnson, for example, will surely locate the border differently than Smith, read the "real" views differently, and so forth.

Smith views ideology as a political disease. Like many others he associates ideology with domination, and simplification with distortion in service of domination. Numerous theorists (e.g., Giddens 1979; Thompson 1984; Habermas 1979) have taken distortion as the essential characteristic of ideology that allows it to function in such a way.[4] Smith agrees: he characterizes ideology as a "device . . . to filter out . . . obviously relevant issues when these issues threaten the status quo," and he criticizes scientists' "predictions that fail to take into account power relations" and thus "hamper our grasp of social reality rather than enhancing it" (Smith 1987: 178, 181).

That ideology can be the consciously and deceptively wielded tool of established interests is evident, but to think that it is only that would be to leave out its role in constituting interests, and to ignore the ideology of nonestablished groups, such as alternative agriculturalists. In short, to define in advance the border between true and false consciousness would seem to limit, prematurely and one-sidedly, the prospects for critical and fruitful dialogue among different viewpoints with differing assessments of truth and falsity.

When positions are perceived as distorted, how can interpretation begin to bridge the differences without either perpetuating deceptions (and thus promoting the views that spawn them) or debunking them (and thus taking sides with the critics)? Some have argued that if we reject a "restricted" or "critical" conception of ideology, then we must either embrace relativism (Bernstein 1976: 106ff) or displace problems of justification "into a murky background where they are conveniently lost from sight" (Thompson 1984: 12-13). Thus it may seem that mediation must succumb to the pressure of radical partisanship: "if you are not for us, then you are against us." If the interpreter debunks *their* tropes then she is for *us*, and if not, then she is against *us*. We propose, however, that the interpreter has another alternative.

The problems generated by the introduction of "ideology talk" are exacerbated versions of the problem we took up at the end of the last chapter—how to find translations that are both fair to those who hold a position and intelligible to those who do not. To this problem is added a layer of suspicion of the motives of other parties (and of the motives of the interpreter) along with an acute sensitivity to the presence of distortion. The solutions remain those we considered earlier. Discourse that is perceived as distorted can be unraveled sufficiently to permit translations that close some of the distance between opposing views. Such interpretation can be reasonably evenhanded if it anticipates critical responses from each side using clues that arise within the disagreement. Interpretation, in other words, can be a means to calibrate conflicting simplifications and patterns of selectivity with one another without positing that some one standard of undistorted communication is necessarily the right one. No doubt some will object to such treatments as merely temporizing moves which the more powerful parties will turn to their advantage. And this may sometimes be so. But it seems unwise to prejudge for any of the parties in a dispute where their advantage lies and whether dialogue will further it. Interpretation, mediation, and negotiation are not the only appropriate responses to conflict, but they are promising ways to draw from disagreements the means for a constructive politics.

Thus far we have introduced a concept of ideology without attempting to answer very many questions about particular ideologies or about what ought to guide our descriptions of them. We have discussed the function and constitution of ideologies only in the most general terms; and we have completely dodged the question of their scope—whether they are all-encompassing or ought to be distinguished from other orientations to thought and action such as science or religion. To many the advantages of ideology talk may still seem to be outweighed by far by its liabilities. It is the burden of the next chapter to indicate why those advantages are compelling.

Notes

1. Freudenberger leans here on two U.N. conferences. One of these finds "deep conflict" in the human environment (Dubos and Ward 1972: 12) and the other makes alarming predictions of the imminent destruction of one-third of the planet's arable soil resources (U.N. Conference on Desertification, 1977). Freudenberger accepts the accuracy of what is claimed in these reports and mounts his appeal on the basis of them. For further development of these views see Freudenberger 1990.

2. Under the influence of structural functionalism in Anglo-American sociology, social scientists have come to associate values either with the idea of a consensus embedded in social institutions or with highly variable, individual attitudes (Giddens 1979). Either values are objective and institutional or they are subjective and individual. However, in controversy the unity of the camps resists being accounted for in either of these ways. Moreover, at least one of the camps we shall encounter finds people related to their culture and society in a way that strongly conflicts with the functionalist dichotomy. So if we lean too heavily on values talk, we risk mystifying the unity of camps and prejudicing the treatment of views which conflict with the functionalist dichotomy.

3. One might ask what then distinguishes the historical consciousness of ideologies from myths and utopias, cultural forms with which ideologies are often confused? Mullins suggests the difference lies in the way events are conceived and presented. Quoting Northrop Frye he argues that in ideology "historical

events . . . are viewed as unique, . . . In mythical time, on the other hand, the concern is not with what is unique in human experience, but with what is universal; 'not with what happened but with what happens . . . in other words [with] the typical or recurring element in action'" (1972: 504). Utopias too present societies characterized by recurrent patterns of change. But they appear as separate from historical process, as finished and perfected, with no expectation that they will ever be realized and, hence with no program for present action (505). Though ideologies may still include myths or ideals of perfection (503), such elements are circumscribed and grounded by the concern to respond to the actual present, something very different from the "universal present" of myth or utopia. Even the traditionalism of someone like Wendell Berry is in this respect quite different from the outlook of a traditional society undisturbed by fundamental changes. The call to recover valued traditions, after all, situates itself in an historically conscious way in the actual present.

4. For example, Giddens maintains that "To analyse the ideological aspects of symbolic orders . . . is to examine how structures of signification are mobilized to legitimate the sectional interests of hegemonic groups" (1979: 188). Thompson pushes the view further to construe the meaning of ideology as constituted by its position in the power structure (1984: 11, 126ff, 173ff). Habermas attempts a general theory of undistorted communication (1979).

5

How to Work with Ideology

Thus far we have been digging through layers of disagreement, trying to find out just how deep they go. They are not just a matter of conflicts over facts or values (or between facts and values) we found; both "facts" and "values" are ambiguous and slippery terms and those who would claim to found their positions in the one or the other are leaving unanswered many questions about the warrant of the positions they take. Beyond "facts" and "values," we explored some competing languages used in agricultural policy controversy. It seemed likely that the particular patterns induced by choices of context and favored idioms might not be intelligible to those not at home with such language. But this level of analysis shed no light on how those choices were made. Finally we took up the possibility that deep disagreement reflects broadly different ways that people see and engage themselves in the world, and we came to regard these frameworks as ideologies.

This sounds all very well, but what is one to do with these ideologies? How, that is, can one describe their role in deep disagreements so as to encourage constructive responses to conflict and hence contribute to its evolution in ways that other forms of inquiry might not? If we are right that ideologies make up the deep stratum of disagreement, their exploration should suggest all manner of possible ways people who deeply disagree might communicate more effectively and even resolve some of their disagreements.

Why Not to Ask Why: Explaining or Understanding?

Having arrived at the deep stratum of ideology our first job must be to cast off the expectation that it will *explain* agricultural disagreement. This is no mean task. Explanation of conflict and controversy is the focus of much scholarly inquiry, and it is certainly tempting to ask why ideologies take the forms they do and why people adopt them. The presence of ideology seems such an obvious and plausible part of the explanation of controversy that one can easily suppose it is the whole explanation. Moreover, causal explanations are looked to so often to provide the means to work with or control situations that one can easily think they are necessary tools for the mediator; it may seem that the whole effort of the mediator should be geared toward explaining *why* people disagree. But there is great reason to be wary of such "why" questions. Indeed, the presumption that people disagree only because of their ideologies may be the most serious source of confusion about our use of the term.

To attempt to explain, particularly to use ideologies to explain disagreements, introduces into the enterprise of mediation two lethal problems. The first arises because explanation is frequently (particularly in heated controversies) a rhetorical means to discredit, dismiss, or debunk an opponent's outlook. Far from enhancing prospects for dialogue, such explanation aggravates tensions. Recall Tony Smith's attack on the role of agricultural scientists at the Iowa State conference on the family farm (Chapter 4). His explanation of their viewpoint as ideological was a means to discredit it. Like Smith, many participants in controversy come in with the view that the outlook of their opponents is already (and all too easily) explained: it reflects gross ignorance, the influence of vested interests, or the like.

Often such explanations are little more than labels. One takes a complex set of images, statements, and actions and labels them as, say, "romantic" or "capitalist." Rather than struggling with the complexities of opposing positions that one neither fully understands nor feels comfortable with, one chooses instead the facile gloss—a convenient short cut to dismissal, a flag to attract supporters: the "romantic's" talk of sublime nature becomes a form of self-hypnosis, while the "capitalist's" litany on progress and productivity becomes a mask for greed. By encapsulating simplified description and loaded evaluation in a single term, one discredits opponents' views and avoids having to engage these views on their merits.

By contrast, the mediator's task is the opposite; it is to invite partici-pants to suspend facile defenses and engage more openly with one another's views by helping them to understand the complex ways in which each view makes sense to itself. If the mediator's interpretations or descriptions are to be helpful, they must scrupulously avoid endors-ing, implying, or presupposing explanations which explain *away* the need to confront that complexity. It is for this reason that we have insisted on "how" questions rather than "why" questions, on investigating *how* people manage to differ in the ways they do (cf Schwarz and Thompson 1990: 90). Although the phrasing of the question alone doesn't guarantee that loaded or reductive explanatory assumptions won't slip in through the back door, it does help to keep description and interpretation sensitive to the fine texture and nuance in the ways people choose to express themselves.

The second problem with attempting to explain in mediation arises from the lack of consensus among the scholars who take up such matters —philosophers, historians, social scientists—as to what comprises an adequate explanation of why people act as they do and express the views they do. Some stress the role of beliefs and the reasons that people have or give for doing what they do, while others stress social structure, political and economic interests, or other factors that might cause people to behave in certain ways even without their realizing it.[1] Associated with each position are distinct views on what needs explaining, what can explain, and what counts as explanation.

That theorists disagree on such fundamental questions is vexing but not lethal. However, in policy disputes, including those about agriculture, the parties frequently appeal to conflicting standards of explanation, and this is a much more troublesome matter. In the family farm conference Berry explained peoples' views and actions in terms of their beliefs and their culture. Boehlje sought statistical correlations to clarify the attributes of different farm structures; while Smith explained the views of agricultural scientists in terms of their place within the power structure.

For the mediator this divergence of standards of explanation is a problem: to endorse any one of the standards is to take sides; it is likely to frustrate dialogue with those who use different standards. A mediator who accepts one of Berry's or Boehlje's explanations, for example, will seem far too indulgent to the other. To avoid thus aggravating the dispute, a determined refusal to take a strong explanatory position would seem to be the mediator's only recourse.

It is these twin problems that drive a wedge between the enterprises of mediating and explaining disagreement. Anything that seems to explain away the complexity of disagreement dismisses the need to understand it. Anything that undertakes to explain why disagreement has come about or why disputants have come to hold the views they do must either take sides or impose a new standard of explanation from outside; either way, it will presume upon, aggravate, frustrate, and further complicate the dispute. But if the mediator must refrain from reductive or dismissive explanations and avoid accounts of why people have come to disagree, what does that leave for her to do? The most important and central task, we have been arguing, is to facilitate better communication and mutual understanding between disputants; and the primary means we have embraced lie in the realm of interpretation or description. As mediators then, the kinds of questions that we most need to address are questions of what the parties to dispute mean by their positions and pronouncements, of *how* they find their own views meaningful. We turn now to the business of shaping these kinds of questions.

The Search for Meaning: Ideologies as Symbolic Networks

Ideologies are the networks of symbols with which we shape our response to the world, understand our place in it, and plan and carry out our actions. They are, in short, the frameworks that provide meaning. But what meanings do they provide and how do they provide them? How, in other words, do they accomplish the business of conferring meaningfulness? The abstruse, philosophical questions of what "meaning" is, of how it can be understood in logic, semantics, and cognitive theory, need not detain us here. From the viewpoint of the practical mediator, we need only note that from a common stock of symbols and language, divergent ideologies have evolved. In shaping the substance of their disagreements people use familiar symbols in different patterns. We can highlight this divergence by beginning with the simplest symbolic units—words or phrases—and proceeding by way of agricultural examples. Along the way we will see how the same symbols vary in significance as they become embedded in different networks.

Much of the conflict (and confusion) in recent agricultural controversy appears to be about what words or phrases mean. An example, and one of the more potent of these symbols is "family farm." As Peterson and others have pointed out, the wide range of uses of the term has been a significant obstacle in making and implementing policies concerned with the survival of the family farm (Peterson et al. 1987). Along with this range of uses, as we saw in the Berry-Boehlje interchange, come great differences in the meaning of the term. They lie at three levels. First, there are differences with regard to the formal definition of "family farm" in agricultural economics, rural sociology, or agricultural policy. There is disagreement over whether a family farm is to be distinguished mainly by the size of the farm, by how it is owned, or by the way it is farmed. The issues at this first level are essentially over what philosophers call the "logical extension" of the term—over which farms belong in the class of things to which the term refers.

It might be hoped that a definition could be offered that would remove this tension, but though definitions have been proposed, the multiple extensions remain. Part of the reason that people are unwilling to give up their uses of the term is that they have commitments to different nonextensional meanings of the term—its "logical intension," the traits or properties that distinguish family farms; or its "connotation," the images and associations it brings to mind (Figure 5.1). A great many terms common to agricultural controversy function in similarly potent ways: "agribusiness," "nature," "productive," "organic," "labor," and "economy" are a few. In some cases terms that are meant to be precise and technical carry a great deal of symbolic force. To some alternative agriculturalists the terms "humus" and "compost" not only refer to particular conditions of decomposing organic matter, but have in addition profound positive connotations, signifying participation in the cycles of creation. "Soil," likewise, is for some far more than the medium in which crops grow, it suggests vitality and bounty. In later chapters we focus closely on a wide selection of such potent terms.

The associations that Americans have attached to the "family farm" are manifold: it has long been a motif in our literature and painting, our theology and political science, and in recent years, our cinema. Some of its associations are profoundly positive: home, warmth, dignity, hard work, commitment, achievement, independence, responsibility, and so on. Some are negative: despair, the destructiveness of nature, loneliness, squalor, overwhelming physical struggle, narrow-mindedness, selfishness, avarice. Both lists could be lengthened, but in our culture these are among the principal attributes that the "family farm" brings to mind.

When we use the term "family farm" we can trust that these connotations are recognized widely enough for the term to resonate in the minds of our audience. At first glance it might seem that defenders of the family farm would rely on positive connotations, while those who see such farming as anachronistic would be apt to draw on the negative images. Yet frequently this is not so. Sophisticated defenders of family farming, like Wendell Berry, accept "despair" and "the destructiveness of

Figure 5.1 Types of Word Meaning

Formal Definitions	– of the class of things to which the term applies correctly	
Nonextensional Meanings	logical intension	– distinguishing traits
	connotation	– associations

nature" as part of the human condition, integrate "loneliness" and "selfishness" into concepts of self-discovery and individuation, and regard narrow-mindedness and avarice as inessential to the mentality of family farmers. On the other hand spokesmen for organizations credited with (or condemned for) the shrinking of the farm sector, such as the American Society of Agricultural Engineers, are not unreceptive to or unmoved by the images of family farm virtues (Stewart 1979). Hence it is not just at the level of intension or connotation that disagreement exists: some who feel the family farm is outdated nevertheless look back on it with genuine respect; many who assert its importance recognize that even in a favorable social setting, family farming is not an easy or always a happy life. Both sides might agree on the connotations the term has in our culture yet still prescribe opposing farm policies.

How can this be so? It would seem odd that advocates of family farming would say it is not easy on mind or body and rarely makes one rich, or that those anxious to move beyond the family farm would look back on it with such fondness. Such ironies have often seemed to partisans to invite a cynical debunking, but they *can* be taken as sincere, as illustrating that the differences involved here run deeper than the meanings of the particular symbols one calls upon, in any of the senses of meaning that we have considered—extension, intension, or connotation. Apparently then, the differences lie at a third level at which terms and phrases, along with their meanings, are situated within much larger systems or constellations of symbols.

Such systems operate as wholes rather than as disjoint collections of the separate symbols that make them up. Even extended figurative or metaphorical themes, such as "the farm as factory" or nature as "the Creation," do not operate independently and do not allow us to distinguish the major outlooks we wish to get at. The theme of the farm as factory is used both by advocates of conventional industrial agriculture and by advocates of a scientifically progressive eco-agriculture, yet it is used quite differently in each case. Something similar happens with "the Creation." Both a progressive and a competing, more humanistic wing of the alternative agriculture movement may regard nature as "the Creation," yet they fit that metaphor into quite different understandings of nature and the human place in it.

To understand how so many disparate symbols can work together within a stable pattern of coherence we need to understand the peculiar links that knit these symbols together, links that involve various kinds of inference, classification, and evaluation. It may help, for example, to consider how people with different outlooks express their differing assessments of the family farm. To some, attachment to the family farm is to be regarded as "romantic" or "nostalgic," and hence "unrealistic." To others family farms allow us to lead "meaningful" and "ecologically responsible" lives and hence are "socially necessary." Both are cases of inference—from "romantic" or "nostalgic" to "unrealistic" on the one hand, and from "meaningful and ecologically responsible" living to "social necessity" on the other. Yet neither customary usage nor formal logic alone can sustain these inferences. One might ask why it is necessarily "unrealistic" to be romantic or why it is necessary to lead a "meaningful" life (whatever that means) and why living on a farm has anything to do with that. But one can only learn what makes these inferences credible to those who find them so by exploring the larger networks of symbols in which they are embedded (see Figure 5.2).

The first of these inferences derives from a way of thinking in which progress is necessary, inevitable, and autonomous. A return to the familiar and beloved ("nostalgia") is not just unwise; progress makes such an attempt futile and illusory; reality demands something different. This way of thinking also gives analysis priority over vision and inspiration, regards the present as determined, and has definite conceptions of how social change can happen and where knowledge and policy must come from. The second inference draws on a different view of nature, human nature, and responsible activity. It assumes that the work we do deeply affects our ability to answer questions such as "Why am I here?" and "What is my life for?" It presupposes that nature is in a steady-state condition that is a bountiful condition as well. Hence to act responsibly is to live in a way that sustains that bountiful state.

Figure 5.2   Types of Symbol System

Informal Systems	metaphorical themes	the farm as factory
	informal inferences	romantic, therefore unrealistic
	images	– determined present – inevitable progress – nature's cycles
	sensibilities	– preferred idiom – preferred ways of acting – preferred source of knowledge
Formal Systems	theories: axiom systems definitional systems, etc.	

Such inferences are affected both by "images"—implicit depictions of what such things as human nature, nature, or society are, can, and should be like—and by "sensibilities"—patterned dispositions to act in or respond to situations in particular ways. To extend our guiding "network" metaphor, we might regard the images as nodes in a network and their meanings as modified both by their connections with other nodes and by their location within the network as a whole. The sensibilities might be thought of as the patterned fields of force set up by the live network, its capacity to do work. The potential functions of a particular constellation of images are thus reflected in the sensibilities it generates or reinforces, and the sensibilities, in turn, are expressed at least partially in the decisions people make and the actions they take.

Neither the constellations of images nor the patterned sensibilities should be thought of as formal systems of belief; they are not philosophical systems or social and political theories. To treat them as such or to identify a specific ideology solely with a particular program of social and political action would both misrepresent the ideology and miss much of its political importance, scope, and flexibility. Unlike formal systems of belief, ideologies are vague, multifaceted, and open-ended. While many people can share the same basic image of, say, human nature or social order, it is quite possible that no two people who share the same image would agree on how precisely to represent it. Nevertheless, they would normally have no problem in recognizing their common outlook or even in acting on it in concert with like-minded others. In forging and sustaining political allegiances and alliances it is these looser meanings—the shared templates and pictured schemata, not the precise substantive claims they may engender—that seem to bring people together. It is as though the images and sensibilities were the stuff of political subcultures always in the making; they are there to be drawn upon, worked with, and shared selectively, but they are never fully articulated, exhausted, or completed.[2]

The deep differences of ideology, then, do not ultimately lie in the meanings of single terms or phrases, or their social and psychological connotations, but in systems of disparate symbols which connect both figurative language and literal concepts and propositions with each other and with prospects for action. Even though they are assembled from materials readily available in the common culture, different ideologies pattern the involvement of people with their society in very different ways. While individuals may be eccentric, or just plain shabby and confused in representing or expressing a particular network of coherence, the patterns of coherence themselves—the ideologies—are not idiosyncratic, but are widely shared, publicly available parts of the social world. For those who would mediate among the ideologies it will be helpful to describe them as wholes and to explore how their components are integrated into a coherent pattern. Only then can we hope to understand why representatives of differing ideologies may define key terms differently, how there can be such ironies in the ways they manage loaded or potent terms, and how even when they agree on the meanings of terms they may still differ greatly in the actions they propose to take.

Mediatory Description as Practical Politics

In the next part of this book we explore the structure of some of these networks in terms of the kinds of questions that ideology helps one answer (see also Chapter 4). We look to find what the various images are, how they work together, and how they lead one to take certain perspectives toward key issues in agricultural controversy.

At bottom our reasons for proceeding thus are practical. As we have tried to show in the preceding pages, the points of view in agricultural disputes are too varied and the points of difference run too deep to allow the mediatory interpreter to presume much. Not much in the way of facts or values will escape being drawn into the vortices of controversy; no one language or favored idiom, no preferred rhetorical strategy will evade contestation; not only will loaded explanatory assumptions and reductive or dismissive explanations aggravate tensions, but even better-grounded attempts to explain disagreement will find competing theories drawn in on different sides of disputes. Nor will mediation make progress by importing standards, frameworks, or concepts from outside of the context of controversy; whatever room to maneuver presents itself within that context, the mediator must accept and work with it or else lose the chance that mediation will be perceived as fair.

The practical consequence of the commitment not to adopt the standards of any one party or to import supposedly neutral standards from outside the controversy is that one must take the statements of the participants seriously. This does not mean reading naively or uncritically, discounting the deliberate use of language to deceive, or the carefully measured use of irony, innuendo, or obfuscation. It does mean being alert to the ideology that is shaping the way the speaker or writer is using language. It means appreciating nuance, inquiring into the reasons for selecting some terms and avoiding others, realizing on a given issue something of the range of images, sensibilities, arguments, and conceptions available in the culture.

A respectful reading of the literature of agricultural controversy is workable partly because the texts one encounters are often rhetorically naive. The strategies of persuasion they use are transparent. But this rhetorical naivete is quite significant; it often reveals how utterly obvious the authors find their outlooks to be and how baffled they are at how to make that obviousness evident to those for whom an opposite perspective is just as obvious. With this obviousness comes the strong emotion that many texts display—a mixture of outrage at those who will not see the stupidity or injustice of their own actions, and celebration of the harmony shared with co-ideologues. The complexity and gravity of the issues amplify these feelings. Yet the bluster and anger, the outrage and self-congratulation, the unexplained precision with which certain terms are used, do make a certain kind of sense—not one lodged in the precision of first principles, but something looser, a rough taxonomy of people and things, a stance regarding history and social change, a view of what human beings are like and what nature is like, expectations about how knowledge is acquired and problems solved, and some notion of an author as a self with a particular role to play within this loose framework.

At times, to be sure, it will be tempting to regard all the discourse we review here—the products of symposia, the commissioned studies, the passionate essays on the problems of our time, the celebratory after-dinner addresses, or the testimony to congressional committees—as so much verbosity (and sometimes pomposity). Yet it is essential that we treat such texts considerately for they are the primary medium of policymaking; the meanings that are put into them and drawn from them are the stuff of policy. In matters of agriculture the policymaker must listen to many voices and draw from many of these texts; will be pressed with considering what is good and just, as well as necessary and expedient; must guess about the future; must integrate the work of many academic disciplines and the views of many partisan groups. The products of policymaking will be documents full of assumptions about people, knowledge, nature, and society, laced with patterns of reasoning, assessments of possibility, expectations of outcome, all of which will be subject from all sides to ongoing interpretation, criticism, and reformulation.

Looked at in one way, what agricultural controversy is about is winning this battle over meanings—meanings of texts, actions, observations of fact, possibilities, recommendations. Consider, for example, some of the familiar questions that shape agricultural controversies (italicized words carry a great deal of the ideological load in each case): does the research of agricultural scientists reinforce the *disproportionate economic power* of the largest *interests* or is it *scale neutral*? Is heavy use of pesticides and chemical fertilizers *mining* the soil, *degrading* the environment and *poisoning* consumers, or is it increasing *productivity* within *manageable* levels of soil erosion, with *marginal* damage to the environment, and *minimal* risks to consumers? Is labeling produce *"organic"* a *"shabby fraud* on a *gullible* and *unsuspecting* public" (CAST 1972, ital ours), or an exercise of the consumer's *right to know*?

These are the kinds of things with which policymakers must deal, the kinds of question they must answer. It will do no good to try to strip the "baggage" of ideology from those terms and phrases, for without this baggage the disputes become vacuous, empty of content, like a fight over nothing. In context each phrase connects to a network of symbols and calls up a framework of images. Each will resonate favorably with some readers and will confuse, frustrate, and challenge others who are more comfortable with different frameworks. Yet authors are rarely explicit about their own images. It is the interpreter's job to infer these—a job that is practically indispensable. To engage in the policy process without a clear understanding of the vocabulary that is being used, the patterns of reasoning that are being employed, the exemplars and images that are giving meaning to what is being said, written, and done is dangerous at best, foolish in the extreme, even absurd. Yet the range and depth of disagreement in current agricultural controversy, coupled with the predominant view that every participant, except perhaps the harried policymaker, should act politically as a partisan of some program (rather than as a facilitator of a process) suggest that a great deal of this preparatory work is disdained and left undone.

Indeed there are many who will bridle at the limitations mediation confronts and eschew the onerous and often thankless work it involves. And there is no shortage of arguments to bolster such rejection. How, after all, can one ever really know that one has located an author's true ideology or the genuine stuff of controversy? Or even that such ideologies really exist and are not just products of an overactive interpretive imagination? Couldn't it be that all the intricate meanings in human intercourse are but masks or smoke screens that cover up the true motives and actual causes, so that, as Giddens observes with regard to accounts given by structural functionalists, the real action always seems to be going on "behind the backs" of the actors? (1979: 71)

Many will remain uncomfortable with the limited and weak tools that alone remain available to the mediator once all the strong ones have been implicated in the dispute at hand, and with the unresolved conceptual, foundational, factual, and normative questions that a sincere attempt at impartiality must live with. Where, after all, do rhetorical posturings end and substantive ideologies begin? What does determine the shape of ideologies? Why do people act the ways they do? What are the right policies for agriculture? To mediate, by practical necessity, is to dodge any of these questions that the parties to dispute do not already agree upon. Indeed, all of these "inadequacies" may be admitted openly without altering the conclusion that mediatory interpretation and description are not only worth undertaking but practically unavoidable in the search for constructive responses to conflict.

In the next six chapters we attempt in a systematic way to supply much of what is central, yet missing in that search. By offering descriptions that are both evenhanded and faithful to the views described, we hope to help shift the dynamics of dispute in agricultural controversy away from strategies of accusation, exclusion, and defense, and toward comprehension, reasoned discussion, tolerance, and cooperation. Such descriptions will not prevent future conflicts, nor will they serve as antidotes for existing conflicts, but they may make it possible to work with conflicts more constructively.

Notes

1. A few have tried with some success to meld the two types of account into one (Scott 1985).

2. Work in progress by Shepard and Dr. Craig K. Harris (Sociology, Michigan State University) documents how these same ideologies appear among commercial farmers in Michigan and affect their choice of farming practices (Shepard and Harris 1989, 1992). It does this through the use of a mail survey instrument that was designed in part on the basis of the descriptions of ideologies that appear below in Part Two.

Describing the Differences

6

Comparing Ideologies

This chapter introduces the next five, which contain descriptions of ideologies as they affect outlooks on agriculture. Here we explain what ideologies we will survey, why we chose them, what works and authors exemplify them, and how the description is organized and carried out.

In this study we examine three ideologies: conventional productivism, ecological progressivism, and radical humanism. We believe these are the main perspectives on contemporary agriculture and agricultural research policy in the United States. Under other labels divisions corresponding roughly to those we make have been recognized by a number of other writers.[1]

These ideologies can be seen as coherent sets of assumptions, simplifications, and expectations with which their users negotiate the gap between real life—"lived existence"—and the orderliness of discourse. They provide categories, significances, and explanations, rationales for undertakings and expectations of their outcomes; they make things make sense.

Classifying Ideologies

Conventional Productivism

Conventional productivism is probably the most familiar of the three ideologies. It has dominated agricultural discourse and policy for at least forty years (Cochrane 1963), but that is not the reason we label it "conventional." Instead, the label reflects the view's content: things will work out without radical intervention, through the continued workings of an historical process (Kaufman 1982: 67; Goldschmidt 1982: 413).

While concepts of historical process have a quite different flavor in Europe and Asia, in the United States the dominant strain of historical-process thinking has its roots in the Enlightenment idea of progress. Having discarded superstition and embraced reason in social, economic, and technical matters, this Enlightenment view advocated a society regulated by the "free market," which would continually improve the human condition.

Such an outlook is common in the literature of agricultural economics and in policy statements emerging from the USDA and from agricultural universities (Douglass 1984: 49). It is held that agricultural policy must freely adapt to changing political, economic, and social conditions. If policy is free to adapt, progress happens automatically. Major farmers organizations as different as the Farm Bureau and the National Farmers Union also embrace conventional productivism. They see a world in which farmers' interests invariably conflict with the interests of other social groups. Farm organizations are therefore necessary to compete effectively and to secure for farmers the best bargains in a policy marketplace (Talbot and Hadwiger 1968; Rohde 1963; Hamilton 1963). Conventional productivism figures centrally in the image we project of ourselves as a nation. We see ourselves as practical, adaptable, energetic; we regard our political and economic systems as adaptive, allowing us continually to meet new needs. The notion of conventional productivism corresponds roughly to what Goulet (1986) calls a "political" rationality and to what Talbot and Hadwiger (1968) call a "Madisonian" ideology.[2]

Ecological Progressivism

As in conventional productivism, solving agricultural problems with new technologies figures centrally in ecological progressivism. But the technologies are not necessarily the same, and the processes which must bring them into existence are conceived quite differently. Innovation is to come not from historical process but from expert social management.

Progressives have long criticized the outcomes of laissez-faire economics and political horse-trading: short-term profits take precedence over long-term needs; alternatives are not adequately compared; a free market cannot assure wise use of resources nor adequate safeguards for the natural environment or human health. As public problems become more complex, the patchwork quilt of compromise becomes increasingly unsatisfactory. Social systems are not self-organizing, but require coherent direction, shrewd guidance, and efficient coordination. Progressives then, are calling not only for better techniques, but for expert management in all aspects of society, as if society itself were a mechanical system. Some progressives see the possibility of an end of history, a state of permanent social well-being that can be attained only if society is properly managed in light of natural and social laws. Frequently progressives will encounter opposition from those who see their management as infringing on inalienable rights of individuals to live as they please.

In Europe ideologies of social management became important in the early modern period among monarchist and mercantilist social thinkers, who saw various forms of central planning as a way to combat the local autonomy of the nobility (King 1949). A strong French technocratic tradition developed in the seventeenth century under the mercantilist administrator J. B. Colbert and survived the French revolution, though with changed personnel and institutions. In England, where local autonomy long remained stronger, a progressive outlook did not become important until the beginning of the nineteenth century when Jeremy Bentham distanced himself from other utilitarian philosophers by recognizing a number of areas of public affairs in which the dynamic adjustments of politics and markets could not guarantee optimal policy (Finer 1972).

By contrast, progressive thinking came relatively late to the United States and it came piecemeal, institution-by-institution. For the most part, early nineteenth century Americans had little tolerance for learned experts, be they doctors, lawyers, soldiers, or financiers. Expertise implied hierarchy, and a great many Americans were mightily suspicious of hierarchy (Talbot and Hadwiger 1968: 22-4). After the Civil War, experts slowly began to be accepted as part of American government. At first they took on tasks no one else was interested in performing—as was the case in the establishment of the federal marine hospitals, the Coast and Geodetic Survey, and the Bureau of Standards. By late in the century such forms of government were moving into more controversial policy arenas: public health, the management of forests, the quality of the food supply. While such administration became respectable in the Progressive Era, it became a pervasive part of government only during the New Deal, when experts were trying to manage everything from plowing schemes to the entire economy.

The progressive outlook entered American agriculture in three main phases. The first was the scientific farming movement of the early and mid-nineteenth century. Like Jefferson, many wealthy American farmers thought of themselves as enlightened and experimental agriculturalists. They learned about chemistry, geology, and horticulture, embarked on plant and animal-breeding programs, experimented with plows and manures, joined learned societies, and wrote journal articles. With the rise of agricultural universities after the Civil War a conflict for authority developed between these gentleman farmers and the professionals at agricultural universities and later at experiment stations, with the latter coming out on top (Marcus 1985).

The second phase began with the conservation movement around the turn of the century. Professional managers like Gifford Pinchot argued that the nation's resources required expert managers to coordinate multiple uses if they were to yield maximum benefit (Hays 1975). In the Country Life Commission, the Farm Bureau, and the extension movement this approach was applied to agriculture; no longer was it assumed that farmers knew best how to farm (Danbom 1979). Roosevelt's 1936 decision to outwit the Supreme Court by linking price-control programs to soil conservation efforts symbolized the expansion of expert management beyond specific resources to the economy as a whole (Halcrow 1977: 158-61).

The third phase began in 1962 with publication of Rachel Carson's *Silent Spring*. By arguing that a business-as-usual approach in agriculture was leading to ecological catastrophe, Carson inspired a generation of agro-ecologists who were concerned that the earth's systems be more carefully managed. Their concerns ranged far beyond songbird extinction: ecological crises (local and global), energy shortages, famine and social disruption in the third world, together with concerns about nuclear power led to a widespread fear that humanity was on the brink of disaster and desperately in need of new and better management (Peters 1980).

It is this Carsonian or ecological phase of progressivism that is most evident in current discussions. It is evident in the concern of many members of the agricultural research community with integrated pest management, energy-saving and low-input technologies, and the importance of biological diversity in farming systems. But it is present also among some researchers outside the agricultural establishment who regard the paradigm currently dominant in Western science as inadequate and hopelessly shortsighted. Some who hold this view seek true knowledge and enlightened agriculture in a quasi-mystical tradition, such as the Biodynamic agriculture movement. Its followers believe that comprehensive technical solutions—virtually panaceas—are available, but do not look for them to come from existing institutions. Elements of ecological progressivism are held by a great many innovative and progressive farmers who have struck a balance between profitability and stability. Many ordinary citizens, unconnected with farming or agricultural research, also endorse the goal of rational and expert administration in government and in the provision of goods and services, including food supply. The ideology corresponds roughly to Goulet's "technical" rationality (1986) and to the "Hamiltonianism" of Talbot and Hadwiger (1968).[3]

Radical Humanism

Radical humanism is likely to be the least familiar of the ideologies we examine here. Although its main themes appear frequently in public policy discussions, they are often overlooked. They may be taken so much for granted that they are not seen either as radical or as part of an ideology or political program. While radical humanists frequently advocate a radical shift away from industrialism, they seek to achieve this aim through a "soft" or cautious means of cultural critique and personal and community responsibility.

Like ecological progressivism, radical humanism is linked to a critique of the industrial revolution. But where progressives tend to worry about the sustainability of large systems and hope to substitute long-term thinking for short-term, radical humanists are more concerned with the effects of industrialization on persons and cultures. In the imperatives of excessive production and consumption, radical humanists find sources of human degradation. With industrialization comes mindless acquisitiveness and passivity in leisure; people lose touch with their capacities for self-direction and self-reliance and become estranged from family and community life. By subjugating skill, dignity, and meaning to a technological imperative to do whatever can be done and make whatever can be made, we are becoming the unwitting victims of progress, not its beneficiaries.

This sort of argument is encountered fairly frequently in the writings of historians, environmental ethicists, and other literati; but the same themes often appear in a more modest form when editorialists or citizen's groups urge a modification of a proposed plant or project on the grounds that it will interfere with the "quality of life."

But if the critique, at least in broad outline, is familiar, less so is the positive side of radical humanism: a set of assumptions about how social and personal decisions ought to be made, about what should guide technical development, about how society can and should change. Common to these assumptions is a view of social change as an outgrowth of changes in the ways of living of a great many people acting independently or in small groups. These will be people who live within or have revived traditions conducive to responsible living. These traditions, in turn, will be embedded in modes of social organization and production that sustain ethical standards and inculcate them into those who grow up in the culture.

Such views are important, if diffuse, parts of the way Americans are accustomed to seeing themselves. We strongly believe in self-determination and self-government. We see our constitution as embodying universal principles of human rights. We are proud of our democracy and extol its responsiveness to "grass roots" concerns. We are confident that our social and educational institutions can regenerate "the American way" in new generations, and act to safeguard the integrity of those institutions.

While elements of both the critical and the positive side of radical humanism are more common than the term "radical" might suggest, the fervent radical humanist will differ from the ordinary "man in the street" in the starkness of her perception of how far we have drifted from our ideals and in her insistence in calling on the present social order to give those ideals more than lip service.

The origins of radical humanism are more complex than those of the other two ideologies. The "humanist" element recalls two sometimes inconsistent aspects of Renaissance humanism—the idea of standards, in aesthetics as well as morality, and the concept of the power, dignity, and centrality of humans, as embodied in the epigram "man is the measure of all things." Central to the Renaissance concept of standards was the idea that balance, harmony, just proportion, and restraint were worthy goals in art and life. The pursuit of unlimited industrial growth or of maximum productivity are clearly inconsistent with this outlook: "enough" does not mean "as much as one can get." Central to the concept of the dignity and power of humans was the view that we have a greater capacity for self-control and self-development than many medieval thinkers had acknowledged. Some writers have found this Renaissance activism responsible for attempts to conquer nature, yet it was equally and has remained an antidote to determinism, a stronghold of the idea that we are capable of emancipation and self-direction and that our own activities can bring them about (Glacken 1967).

Humanistic criticisms of industrialism emerged in the nineteenth century and drew both on classical and romantic traditions. One tradition, drawing mainly on the theme of aesthetic and moral standards, begins with Carlyle and runs on through Ruskin, William Morris, and the pre-Raphaelites. Another draws on the theme of enlightenment of the romantics and transcendentalists, with Goethe and Shelley as especially articulate spokesmen. There were other sources: eighteenth-century rationalism (Jefferson) and various evangelical movements (Lord Shaftesbury, Leo Tolstoy, or numerous Quaker reformers).

The chief political philosophy embodying these themes in the nineteenth century was non-violent anarchism, though anarchism was itself not one but many movements. Anarchists believed that ordinary people living in communities where education for responsible living was practiced did not need government. They tended to support direct as opposed to representative democracy, decentralized institutions where institutions were necessary, and hoped to develop cooperative barter economies (Ward 1974).

These movements were not necessarily radical. Some American anarchists placed themselves in the tradition of the Declaration of Independence and many of their concrete concerns—regional autonomy, cooperative endeavor, direct democracy, concern for local control of education—became part of the American political and cultural scene without ever being labeled anarchistic (Lehning 1973).

A great many of these themes, both of humanism and anarchism, were tied together in the 1930s in Lewis Mumford's *Technics and Civilization*, and in similar syntheses by others: critics of industrialization such as Ralph Borsodi and Stuart Chase, Marxist back-to-the-landers such as Helen and Scott Nearing, the distributist movement, the Catholic Rural Life Movement, and the Southern agrarians. In the 1960s and 1970s many of these themes were adopted in the youth counterculture, by those influenced by humanistic psychology with its emphasis on growth and maturation, and by others both on the left and the right, who were critical of the material, cultural, and spiritual control imposed on them by society (Shi 1985; Peters 1980; Shapiro 1972; Roszak 1972).

In recent American agricultural controversy the radical humanist outlook shows up in a number of ways. Radical humanists take particularly seriously the loss of the family farm, which seems to them especially well suited to foster self-reliance, flexibility, practicality, creativity, humility, restraint, and cooperation. They are concerned that the export of western agricultural technology not disrupt indigenous cultures, and they tend to elevate self-determination over greater productivity or material well-being. They are also concerned to establish local control over marketing and production (economic democracy).

Though radical humanism is rightly associated with alternative agriculture, it is present too in the agricultural research establishment, though often subordinated to other ideologies. The proportion of farm-raised agricultural scientists remains high; and as the recent anniversary history of the American Society of Agricultural Engineers makes clear, there remains a great deal of ambivalence about the disappearance of the agrarian life (Stewart 1979: 357-8). Concern for the social and cultural effects of the technologies of industrial agriculture is often expressed, even if rarely translated into actions that satisfy radical humanist critics. Our notion of radical humanism corresponds roughly to what Goulet (1986) calls an "ethical" rationality, and to what Talbot and Hadwiger label a "Jeffersonian" ideology (1968).[4]

A few comments on these ideologies are in order. Most writers who take part in agricultural controversy clearly belong to one of these frameworks, but not all do—some have unusual assumptions that may perhaps reflect a European or Asian heritage. Others appear genuinely confused—one page seems incompatible with the next. Others may deftly blend or change frameworks, suiting their writings and even their actions to the occasion, perhaps to forge a political alliance. It is also important to recognize that these ideologies are not permanent and immutable, but continually evolve under the pressure of ideological competition and changing circumstance. It is also wrong to think that they represent mutually exclusive programs for agriculture and that we must follow one rather than another. The images, assumptions, and modes of acting that make them up will doubtless shift in the future; intractable oppositions may give way to alliances. Indeed, possibilities for accommodation or alliance provide one motive for this book.

The range of current understandings of "agricultural sustainability" illustrates how agricultural controversy can be seen as a conflict of these ideologies.

Ideologies of "Sustainability"

In his introduction to the 1984 volume, *Agricultural Sustainability in a Changing World Order*, Gordon K. Douglass identified three images of sustainability held by three groups of writers on agriculture. First were "defenders of Western methods of farming—of highly specialized, mechanized, chemical-intensive, science-based methods applied to increasingly large farm and distribution units." For these writers (our conventional productivists) "sustainability" meant providing food to sustain the human population at whatever level it exists. Achieving sustainable agriculture involves predicting future food needs and finding economical means of meeting them. Like all our ideologies, this conventional view is coherent, ethical, and plausible. Its coherence comes from the perception that dealing with the problem in the short term is the way to secure the future (or to paraphrase an old saying, "take care of the years and the centuries will take care of themselves"). It is ethical in that it accepts responsibility for the food needs of all people. Its plausibility is drawn from historical experience: in the past, free market production, aided by science and technology, has expanded to meet increasing demand and at the same time lowered the cost of food to the consumer.

Douglass' second group of writers (corresponding to our ecological progressives) is united by a concern for "ecological balance rather than economic scarcity." He divides this group into two subgroups, one which sees ecological balance as possible with only "marginal adjustments" to conventional agriculture, a second which sees need "for a radical restructuring of agriculture . . . since their reading of the ecological evidence . . . and of historic experience, is apocalyptic" (5). What is to be sustained in this case is the stability of a biophysical system that permits a substantial level of food production into the indefinite future. In this view population is not an independent variable; population control is a central part of the problem. Here "sustainability" presumes stability, and so long as one element in the agricultural system—population—is highly unstable, there can be no sustainable agriculture. Since population stability will only be achieved slowly, this view emphasizes cautious use of nonrenewable resources and protection of the systems which provide renewable resources. The coherence of this view rests on the viability of a balanced ecosystem. It is ethical in acknowledging and seeking to secure the rights of future generations. Its plausibility is drawn from basic physical laws, such as the laws of thermodynamics and the conservation of matter, which it promotes to regulative status. No matter how much we pay, we cannot escape these laws, it is argued, so we would be wise to manage our affairs in accord with them.

Douglass's third group of writers is concerned with sustaining communities, human dignity, and freedom, and corresponds to our radical humanists. They see modes of production as affecting culture and social organization, and are searching for an agriculture that enhances community, individual freedom, perceptions of self-worth, and social equality—usually a system in which food production is largely in the hands of many small proprietors. As Douglass notes, many who take such views are concerned also with the sustainability of the biophysical systems upon which communities depend, yet they are likely to focus on these problems at a local level. In this view global agricultural sustainability will be achieved when problems are solved in individual communities. What gives this view its coherence is the proposition that traditions of responsible living thrive in certain social and cultural environments. It is ethical in seeking to empower people to shape their own lives. What gives this view its plausibility is the record of responsible grass roots action and the existence of stable and viable communities.

What for Douglass was a way of making sense of the range of ways writers, mostly academics, conceived "sustainability" has since become a matter of immediate political and economic import. A resurgence of global environmental concern in the late 1980s, and publication of the National Research Council's report on *Alternative Agriculture* (NRC: 1989) have placed "sustainablity" high on the agenda for agricultural technologies, policies, and research programs. A flurry of articles in the trade and academic press reveals the contest over what "sustainability" is and who will speak for it. A writer on the op-ed page of *BioScience*, for example, complains that a human food supply based on genetically-engineered "ligno-cellulosic" perennial plants, however limitless it might seem, would be but "a counterfeit version of . . . sustainability"—such an industrial agriculture would fail to sustain rural communities and meaningful lives (Orr 1988). Elsewhere, writing under the title "Planetary Patriotism," two Monsanto vice-presidents insist that "sustainable agriculture is possible only with the biotechnology and imaginative chemistry"; traditional, low-input approaches will sustain neither the growing human population nor the competitiveness of the American farmer (Schneiderman and Carpenter 1990; see also Beus and Dunlap 1990; Buttel and Gillespie 1988; Dahlberg 1991; DeLind 1991; U.S. Congress 1990).

Such an example suggests the uniqueness of each outlook; it makes clear that one needs to go beneath assertions of the importance of "sustainability" and look at what is to be sustained, how it is to be sustained, and why. Below we introduce the central texts and authors we have chosen to illustrate these ideologies more fully.

Exemplary Authors and Works

Wendell Berry

We use the writings of the Kentucky farmer-critic-poet Wendell Berry to exemplify the radical humanist ideology in contemporary agricultural controversy. Berry farms 75 acres of hillside in the Kentucky river valley, the area in which he grew up, and also teaches English at the University of Kentucky. In 1977 he published *The Unsettling of America: Culture and Agriculture (UA)*, an expansion of a 1972 essay, "Discipline and Hope," where he examined alternatives to the prevailing "general cultural disorder" (*D&H*: 86).

The Unsettling was widely read and the book confused many readers. Some chapters struck familiar themes: the crises of big technology and environmental despoliation, the solutions of appropriate technology and alternative agriculture. Others—especially those on character, culture, and one enigmatically titled "The Body and the Earth"—were more perplexing. As some critics recognized, Berry had responded seriously to the question of what sort of culture a renewed agrarian society entailed. Unlike many of the youthful utopians of the 1960s and 1970s who idealized life on the small farm, Berry made much of the frustrations and sacrifices farming involved. He wrote about the sort of mentality and culture that allowed farmers not only to endure their way of life but to embrace it and be enriched by it.

A key concept in the book was the idea of settlement. By "unsettling" Berry did not mean that America was being depopulated; "settlement" signified a settledness in a place, a comfort with the life one leads there, and one's hopes for the lives one's children will lead there. The way to such a state of mind was through a revival of culture that would bring significance and sacredness to ordinary life. Such a culture would recognize the necessity of restraint and would regenerate a delight in living within biological limits rather than trying always to exceed them. Because it would bring with it a closeness to the land and a respect for nature, such a culture would lead also to ecologically sustainable agriculture.

Besides dealing with these philosophical, psychological, and even spiritual themes, *The Unsettling* and subsequent essays also brought out the political, scientific, social, and practical implications of Berry's perspective. He argued for a truly interdisciplinary agricultural science in which problems were formulated in terms of particular "whole" farms, to be understood as unique and integrated patternings of traditions with practices, enterprises, resources, and skills. He saw great need for social change but argued that the living of individual lives was a far more powerful social force than were laws or policies, lobbies or protests. He opposed directions in technological development that failed to acknowledge natural limits or human needs, and he rejected the concept of government as a provider of services and satisfier (and manipulator) of constituencies.

Berry's intellectual roots are eclectic. He has been seen as a disciple of pantheistic Emersonian transcendentalism, Jeffersonian agrarianism, or modern paganism (Hicks 1979; Pevear 1982; Lang 1983).[5] But as T.D. Young has suggested, he is in many ways reviving the crusade of the Nashville agrarians of the 1920s and 1930s (1978). In 1930 John Crowe Ransom, Robert Penn Warren and ten other southern men of letters published *I'll Take My Stand*, an anthology which articulated an agrarian alternative to industrialization in the south. As would Berry, they saw need to turn away from the unending worship of progress and toward a way of living in which culture supplied meaning and restraint. Politically, like Berry, they argued for "distributist" land policies which would put land ownership in the hands of many, and make farming the main way of life. The movement was part of a decentralist and distributist coalition in the 1930s which included those influenced by the English distributists, G. K. Chesterton and Hilaire Belloc, and the Catholic Rural Life Movement (Shapiro 1972). A second source, as Young recognizes, was the "new humanism" of Irving Babbitt and Paul Elmer More. Especially in the work of Babbitt, this literary and social movement drew upon Buddhist ethical teachings to develop concepts of balance, harmony, and restraint in both art and life.

Berry showed the cogency of these ideas to the prevailing environmentalism, the back-to-the-land movement, and the counterculture. In their opposition to industrial capitalism, and their pursuit of peace, balance, and meaningfulness, and attraction to eastern religious traditions, these modern movements echoed the concerns raised by the critics of the 1920s and 1930s. Our examination of Berry's works focuses on *The Unsettling of America* and "Discipline and Hope," using his more recent essays, especially in the Gift of Good Land (*GGL*) and Home Economics (*HE*), for clarification.

Charles Walters jr

The writings of Charles Walters jr illustrate ecological progressivism. They reflect a blend of alternative agriculture and radical agrarianism (in contrast with the literary and philosophical agrarianism of Berry)—the heritage of the Grange, the Populist Alliance, the Farmer's Holiday Movement, the National Farmers Organization and similar organizations down to the present. These farmer-solidarity movements were to give farmers more political and economic clout than they could wield as individuals. One of their common objectives, control of the institutions of agricultural science, was seen as crucial in achieving satisfactory agricultural policy (Salutos and Hicks 1951: 128, 229-30). Like many critics of big agriculture they saw many of the problems of farming as a result of corrupt science working in service of big business.

Walters himself was deeply involved in the National Farmers Organization during its militant years in the late 1960s. Two of his early books, *Holding Action (HA)* and *Angry Testament (AT)*, recorded and defended the NFO's milk and meat holding actions; a third, *Unforgiven (U)*, dealt with its economic philosophy. This was centered on a physiocratic "multiplier principle" developed during the 1930s by Carl Wilken and the Raw Materials National Advisory Board. Wilken and Walters held that since farmers' wealth came directly from the soil, it was "new" wealth. Through farmers' expenditure wealth was multiplied throughout the economy. Hence strong farm prices were the key to prosperity.

As NFO activism declined during the early 1970s, Walters shifted to ecological aspects of agriculture. *The Case for Eco-Agriculture (CEA)*, a short work, appeared in 1975 and the monthly tabloid, *Acres U.S.A., The Voice for Eco-Agriculture (Acres)*, began in 1971. Avoiding countercultural concerns—communes, meaningful relationships with the earth—Walters concentrated on novel fertilization and pest control techniques suitable for commercial farms, on sophisticated technology for diagnosing subtle (and controversial) forms of chronic plant malnutrition, and on heterodox scientific theories which explained what was wrong with our food, our farm practice, and ourselves. Orthodox agricultural science, owing to its connection with big oil companies and other powerful institutions, had neglected the true complexity of agriculture and nutrition, Walters wrote. In *Acres* he took up the causes of research programs rejected by conventional science. He championed the University of Missouri agronomist W.A. Albrecht, who between 1918 and 1952 had developed a far more sophisticated physio-chemical approach to soil fertility than the NPK (nitrogen, phosphorus, potassium) approach of conventional agriculture. He publicized the biodynamic farming of the German mystic Rudolph Steiner, in which "etheric" forces unrecognized by conventional physical science were believed to affect plant growth and health significantly.

For Walters, "eco-agriculture" represented not so much the rejection of agrarian activism as its maturation. The unorthodox economic views of the NFO years continued to appear regularly in *Acres U.S.A.* He continued to appeal to the better educated and more freethinking farmers who had supported the NFO. These farmers would recognize the impossibility of the economic and biological situation in which they were working and see the need for a radical alternative. Indeed, as we shall see, "eco-agriculture" both explained the NFO's failure and offered the only possible route for achieving its goals. Our analysis of Walters' views comes both from his political and economic works—*Unforgiven, Holding Action,* and *Angry Testament*—and his eco-agricultural works—*Acres U.S.A., The Case for Eco-Agriculture,* and *An Acres USA Primer (AP)* (coauthored by C.J. Fenzau).

The Battelle Memorial Institute's Agriculture 2000

Agriculture 2000: A Look at the Future (Battelle 1983a) is one of many assessments of future agriculture to appear in the last 20 years and it, like most others, foresees a future of increased productivity, more sophisticated technology, and fewer but larger farms. In contrast to the works of both Berry and Walters, who are concerned with what agriculture should be like in the future, these reports deal with what agriculture *will* be like, if present trends continue. Although one can find aspects of this approach in the social surveys of the 1920s or in the earlier Country Life Commission, the prototype of the current strain of agricultural futurology was the 1962 report, "An Adaptive Program for Agriculture" of the Committee for Economic Development, a group of economists and business leaders who recognized that new technology would reduce the proportion of the population that made its living in commercial agriculture (CED 1962). Such a view had influenced the policy of Eisenhower's Secretary of Agriculture, Ezra Taft Benson, and continued to influence subsequent Democratic administrations (*AT*: 32-5; Brooks 1982). The theme of these reports, and of *Agriculture 2000*, is that American

American agriculture is on the right track, the most rational track, and perhaps the only possible track. The first thing one reads in the foreword to *Agriculture 2000* (by former Secretary of Agriculture John Block) is that "Agriculture in America is a success story" (ix). The miracles we have seen are but a hint of what is to come with the perfection of monitoring and control instrumentation and biochemical and genetic intervention.

The report we consider here, *Agriculture 2000*, was researched and written by the staff of the Columbus Division of the Battelle Memorial Institute, one of the nation's more prestigious economics-and-technology think tanks. It was sponsored as a fiftieth birthday celebration by the Production Credit Associations, the federally sponsored local banks that make loans to farmers, and one of the few New Deal agricultural programs to survive. It is common for institutions to mark an anniversary with an institutional history. That the PCAs chose a report on the future is significant, a reflection of confidence in the status quo—so firmly are we in control and so morally right and appropriate is that control that we may congratulate ourselves not just on what our policy has done but on what it will do.

The book itself is a 180-page summary of a more detailed technical report which we have also consulted. This summary report consists of ten chapters in two main sections. The first section on "Trends in Agriculture" includes chapters on "Current Agricultural Trends," "Regional Agricultural Characteristics," "Developing Agricultural Technologies," "Marketing," "Agricultural Inputs and Finance," and "Quality of Rural Life in the Year 2000." The second section, "Research and Technological Development," has chapters on "Crop Technologies," "Mechanization Technologies," "Technologies in Animal Agriculture," and "Communications and Information Management." To the degree that the particular concerns of the PCAs can be identified in this book, it is in its emphasis on financial aspects, both with respect to the financing of farming and the marketing of farm products.

It would be wrong to think that these authors and texts represent every kind of literature, every social context, or all the intellectual roots that can be associated with each camp. Nor should these texts be looked at as manifestos of shared beliefs. Some radical humanists will find Berry too literary, too insistent on universal values, or too unregenerate an agrarian. Some ecological progressives will find Walters parochial in his defense of the Midwestern farmer, or will reject his political and economic programs, or deny the validity of his brand of eco-agricultural science, while still agreeing that scientific agriculture is essential.[6] Some conventional productivists will find the Battelle report simplistic and sensationalistic, or may disagree with its projections, or be concerned that its arguments on ethical and social issues are undeveloped.

Indeed, the very idea that there is or should be an orthodox version of each of these camp's position is misconceived. The unity of each camp is not so much a unity of belief as a common attraction for certain themes, images, metaphors, and other means of orienting thought and shaping engagement. These constitute meanings and lead to beliefs. What the three works-authors do, is to illustrate three very distinct ways of understanding and interacting with the world of agriculture. Each has the depth, breadth, and coherence that makes it possible to lay open for examination their orienting themes and images.

How to Compare

The next five chapters compare the three authors with respect to their images of human nature, nature, knowledge, social order, and responsible action or praxis. Here we explain why we use these categories of images, what each category includes, and how we extract examples of each sort of image from texts.

Earlier we developed the idea that a concept of agriculture is the product of an ideology applied to the policy issues, institutions, and practical problems that bear on the production and distribution of food. If we take "technology" to refer to a system of knowledge of how to produce goods and services (not just to techniques themselves), each of these agricultures is a technology—each offers a systematic understanding of the proper way to produce and distribute food. What gives each agriculture its coherence are the implicit assumptions that make its program seem sensible, proper, necessary, or good. The categories we use here organize these assumptions and highlight that coherence.

We start with *images of human nature*. People cannot avoid making assumptions about human nature when they claim that a technology or an agriculture will work; technologies are to serve people, and their implementation—at all levels from making long-term policy commitments to plowing fields—must be a human achievement. To raise anew complex questions about what human nature is each time a decision has to be made to set price supports or start a new agronomic research project is surely a waste of time; hence we go comfortably on, holding and using assumptions about human nature that we have long held, feeling sure that our own experiences have made us expert enough.

Yet the things we do may not make sense outside the context of our own views of human nature. Included in any such view are assumptions about what people like or can be persuaded to like, about what is good for us and others, about how responsible people are and what they are likely to be responsible to, about whether they can be relied upon to know what's good for them. In many cases agricultural controversy is, in a direct sense, a disagreement over such assumptions. Some, like Berry, argue that people really like to farm and that farming is a good way for humans to live that has been made unattractive by current agricultural policy. Others, like the agricultural economist Luther Tweeten, believe that humans do not especially like to farm and that it is not a uniquely valuable thing for them to do (Tweeten 1983). Each side may cite opinion polls and sociological studies to uphold its claim, but to no avail, for the positions reflect assumptions not only about what people want, but about whether they really know what they want and what is good for them.

Images of nature are equally unavoidable, though frequently more explicit than images of human nature. It is from nature that we get food; some image of nature thus takes part whenever we discuss agriculture. Take the important question: will nature continue to feed us? Some believe nature can always be adapted to meet our needs, if we try hard enough. To others, nature is only narrowly adaptable; we must discover its narrow limits and learn to live within them. Still others find nature powerful, awesome, and magnificent, and see great harm in trying to push it to its limits. Better science will not eliminate these divergent views; they exist despite all the studies of ecologists, climatologists, agronomists, and engineers. Besides this big issue of what we can expect nature to do to or for us, agricultural controversy also takes in other questions about nature—whether it has a purpose and what that is, whether it is a unified entity or a conglomerate of matter and energy conveniently lumped under a single term, how and to what degree one can know nature, and what the relationship is between nature and human nature (see Bramwell 1989).

Images of knowledge form the third category. Each ideology supports itself on the basis of views about what can be known, how it can be known, what sorts of people are likely to know it, and what the knowledge is to be good for, i.e., who it should be communicated to and how it should figure in decision making (Elkana 1981). These views manifest themselves clearly in longstanding debates about the purpose and organization of agricultural universities and about the priority of general scientific knowledge over the complex, site- and context-specific knowledge farmers need. Sometimes farmers have insisted that agriculture professors be farmers themselves to insure that the professors produce practical knowledge (Salutos and Hicks 1951: 437).

Like images of human nature, *images of social order* may seem to belong to the context in which an agriculture operates, rather than to agriculture proper. Yet such images are nonetheless essential to the coherence of each agriculture. Our images of social order are reflected in how we classify social phenomena, whether in terms of nations, classes, cultures, religions, occupations, economic systems, genders, ages, or regions. Any of these can be the focus of an analysis of how agriculture is or should be integrated into society, but most writers see one, or a few of these units as critical to social organization. Berry, for example, focuses on the marriage as the essential biological, and hence social and cultural unit. The fundamental conflicts occur within the person, the marriage, and the family. Conflicts at other levels are illusory or unnecessary and often reflect a failure to resolve a more fundamental conflict. Walters, on the other hand, sees humanity as divided by occupation, geographic situation, and class. Because of these divisions there will invariably be conflicts of interest: farmers will ever be in conflict with consumers, cities with rural areas, and landowners with tenant farmers. For the Battelle report the primary distinctions are occupations, economic systems, and nations.

As with the other sets of images, it is futile to hope to resolve these divergent views through a sociological analysis that would show one of these schemes as truly primary. Even if it were proved, say, that occupation was the strongest determinant of behavior and hence the most important basis for distinguishing groups, it could still be (and is) argued that such is only the case in the imperfect society of the present, while agricultural policy should be geared toward creating the sort of society we "really" want or that ought to be.

This matter of planning—whether for radical change or for the continuation of what we have today—leads directly to the last category, *images of praxis*, or effective and responsible social action. Central to each ideology is an issue of change: whether the changes that need to happen require a sharp break with the past or whether they will come about automatically through the perpetuation of present behaviors, roles, and institutions. Closely connected are questions of who is to be responsible for taking action, when it is necessary, and what the modes and limits of action are. Included here are familiar ethical questions—the rights of farm animals, the justness of mechanization research, the wisdom of depleting groundwater for irrigation. Images of praxis provide ethical standards and at the same time they depict the practical context of effective, responsible action. Without images of praxis, which in turn are linked to images of human nature, nature, knowledge, and social order, we would be at a loss to say how ethical claims bear on actual conduct.

In the descriptions that follow we hope to represent the three agricultural ideologies as coherent, alternative approaches to agriculture and agricultural policymaking. We have tried to ask the same questions of each of them, to elicit answers in depth, and to make clear the way in which each makes sense in its own terms. Readers may also want to note that the clearest view of the central organizing theme of each ideology appears in a different chapter. The core of Berry's radical humanism is in his conception of human nature. The key to Walters' outlook is developed in the chapter on images of nature; the heart of the Battelle report's outlook is only developed in the final chapter on images of praxis.

Notes

1. These include Copeland 1980; Douglass 1984; Marien 1977; Rautenstraus 1980; Stockdale 1980; Coleman 1982; Mitroff 1974; Goulet 1986; Youngberg 1978a,b; Miller 1984-5; Schwarz and Thompson 1990.

2. Among those who have represented this point of view in matters of agriculture and agricultural research policy are Emery Castle (1978), Ross Talbot and Don Hadwiger (1968), Luther Tweeten (1978), Willard Cochrane (1963, 1979), Kenneth Boulding (1963), Boysie Day (1978), R.C. Rautenstraus (1980), and Pierre Crosson (1982: 5).

3. Among those who have taken an ecological progressive position in recent writing on agriculture and agricultural research are Thomas Edens and Herman Koenig (1980), W. R. Raymond (1980), Dietrich Knorr (1981), Eliot Coleman (1982), George Boody (1982), Michael Perelman (1982), John Todd (1984), Wes Jackson (1984), Anson Bertrand (1980), C. A. Black (1978), Kenneth Dahlberg (1979), Richard Harwood (1984), Garth Youngberg (1978a), William Lockeretz (1983), M. Kiley-Worthington (1981), some of the writings of Robert Rodale (1972), and Marty Bender (1984).

4. Among those who represent this outlook in matters of agriculture and agricultural research policy are Gene Logsdon (1984), Donald Worster (1984a), Dana Jackson (1984), Angus Wright (1984), Gary Paul Nabhan (1984), Gary Snyder (1984), Robert Rodale (1972), Lawrence Busch (1984b), Stillman Bradfield (1981), Lawrence Goodwyn (1978), Maynard Kaufman (1982), Walter Goldschmidt (1982), John Shover (1976), Sara Ebenreck (1982), and Jerry Moles (1982).

5. For other views of Berry's importance see Campbell 1990 and Montmarquet 1989: 235-40.

6. Of the three, only our choice of Walters has elicited strong criticism. Some early reviewers of the work in progress saw Walters as marginal. This reaction is not surprising. Within the agricultural establishment Walters is widely regarded as a kook, as were the Rodales for decades. One reason for this reaction is that attacks on academia are central in the populism Walters espouses; he stands apart from academic traditions, just as his "true scientist" must be independent from "the powers that be." But however marginal such a view may

seem within the academy, it does resonate with a significant population (Schneider 1987). Moreover, the theme of criticizing academics is so common in ideological discourse (even construed nonpejoratively) that the question here is really not *whether* full-blooded American ideologies criticize academics, but *how*. Berry and the Battelle report also engage in this ideological sport, though in different ways from each other and from Walters. A second reason for concern about our use of Walters appears to be rooted in the fact that some of the scientific claims he endorses transgress the bounds of scientific orthodoxy. This issue too, however, is a longstanding characteristic of agricultural controversy and cannot be arbitrarily labeled "out of court" by a mediator. Finally, in addressing both the social and the scientific implications of an ecological progressive outlook, Walters has gone further in articulating what we regard as a common outlook than have most others. It is for this reason that we find the choice of Walters and the resulting scrutiny warranted.

7

Agriculture and Human Nature

Technologies don't make sense without humans. Skills and machines are creations and possessions of humans; they are used toward human ends, and indeed to a far greater range of ends than are usually contemplated in technology assessments. With respect to agricultural technologies, it is clearly important to have a conception of what farmers, consumers, or agribusiness people are like, or in the words of Geertz "what the devil they think they're up to" (1979: 228). But it is also important to recognize deeper images of humanness that are not limited to catalogs of attributes of those in particular occupations or cultures. A good deal of the disagreement about the direction of agricultural research is grounded in differing assessments of human nature—of human desire, e.g., what farmers want farm life to be like (and what they ought to want farm life to be like); of human potential, e.g., how quickly people learn and from what sources they learn; of human character, e.g., how responsible people are likely to be in their economic behavior and what they are likely to be responsible to or for; and of the human condition, e.g., the conditions and limits within which moral behavior is possible.

Some may argue that these are questions for the empirical research of sociologists and social psychologists. Even if this is so, influential writings on agriculture continue to embody conceptions of humanness which are not drawn from full or critical acquaintance with the findings of social scientists. And empirical findings about human nature can be interpreted in very different ways. The images of human nature held by our three writers will often lead to different interpretations of the same empirical data.

Our survey of images of human nature focuses on three issues: (1) the nature of humans—what are humans like and what difference does it make what they're like; (2) hope—what should (could) they be like and how will they get that way; and (3) freedom—in what sense are humans

free and what will they do with their freedom. Not all these questions
are fully answered in this chapter, but we do begin the job of setting
forth the divergent frameworks within which each ideology answers
them.

Berry

Berry—The Nature of Humans

The main problem that concerns Berry with respect to human nature
is why so many people in industrialized countries are bewildered,
distrustful of others, cut off from communities, dissatisfied with their
own lot, but at a loss about where to turn. How, he wonders, did we get
this way, and how are we to find a way out of it. He wonders also about
the relation between this pervasive dissatisfaction and contemporary
social and environmental problems, such as the destruction of farm
communities and the loss of farm land. For Berry, as for many critics of
agriculture, one addresses these problems in terms of the biological
realities of human existence—birth and death, growing food and eating
it, reproducing the species. Continually he returns to these realities,
asking us to acknowledge them, to think about what stance we will take
towards the limits they set and how we are to enhance our humanness
without denying our insignificance in a vast universe.

These issues are central for Berry. Because he does not see nature as
designed for humans, he is pressed with much the same problem that
faced Darwin and Thomas Henry Huxley in the second half of the
nineteenth century: how it is that a conscious creature in a world of
instincts and accidents could ever produce a just and stable society of
morally aware and responsible individuals. To the degree that humans
are like animals, struggling to survive and reproduce, succeeding by
"brute strength [and] cunning, . . . [regardless] of value and of consequence" (*UA*: 122), this seems inexplicable. In the Western tradition there
have been two broad approaches to resolving this problem. Some have
derived these peculiarly human attributes materialistically, as adaptations
of a society to conditions of life. Others have posited some sort of soul,
a non-material, moral sensibility uniquely human. Berry is uncomfortable
both with dualism and materialism; he finds a third explanation in a
concept of cultural health. We do have a wildness, he admits; his use of
"instinct" or "human nature" refers to undisciplined, even violent
tendencies that exist in us all. Humans do not necessarily act responsibly, are often undisciplined in their sexuality, their acquisition of food,
or exploitation of land, and are unwilling to do necessary work (*UA*: 105-

6, 113, 119, 130-1, 135): "Human nature is such that if we waited to do anything until we felt like it, we would do very little at the start, even of those things that give us pleasure, and would do less and less as time went on" (*D&H*: 117).

But humans also have the capacity to create culture, and that is our most important attribute, for culture can bring discipline and restraint, a morality that is our means of surviving in the natural world: "a live and adequate morality is an accurate perception of the order of things, and of humanity's place in it. By clarifying the human limits, morality tells us what we risk when we forsake the human to behave like false gods or like animals" (*D&H*: 166). Thus culture is not an effete matter of interest only to moral philosophers but a practical concern of everyone:

> A culture is not a collection of relics or ornaments, but a practical necessity, and its corruption invokes calamity. A healthy culture is a communal order of memory, insight, value, work, conviviality, reverence, aspiration. It reveals the human necessities and the human limits. It clarifies our inescapable bonds to the earth and to each other. It assures that the necessary restraints are observed, that the necessary work is done, and that it is done well (*UA*: 43).

But the modern world is full of people questioning their cultures, testing the limits and possibilities that their tradition has laid out and striving to move beyond them by pushing back the boundaries of possibility with the help of new and powerful technologies.

To Berry this tendency is at the center of the modern malaise; both the deep dissatisfaction that pervades modern living and the seemingly inadvertent destructiveness of modern technological expansion come from our failure to live within a culture: "our bewilderment is not the time but our character. We have come to expect too much from outside ourselves" (*D&H*: 167). Lured by promises of a better future through new technologies and befuddled by the abstractions of a scientific way of understanding, we have cast ourselves adrift from the traditional moorings of meaningful living, eschewed the practical need to continue to transmit the insights of the past, and forgotten that our own character is the central problem and inescapable project of human existence. We must learn, says Berry, from the Confucian aphorism of the "archer, [who] when he misses the bullseye, turns and seeks the cause of the error in himself" (*D&H*: 168, cf "The Doctrine of the Mean," Chan 1963: 102).

We would be wrong here to think of culture and instinct (or morality and human nature) as opposites; to think that Berry is referring like some Puritan to an inner struggle between beast and saint. For Berry, this would be a false dualism, resting on a misunderstanding of how cultures work and what is meant by accepting necessary disciplines. Culture, here, is not the external hand that represses innate cravings; that we might see it this way is for Berry a further indication of the disorder into which our own culture has fallen.

But the concepts of culture and "the larger disciplines" are also hard to understand because they are closely tied to the Confucian ethical tradition which Berry relies on to adapt a core of Christian humanism to the problems of the modern world. The concepts warrant some development for they will figure centrally in Berry's perspective toward nature, knowledge, social order, and praxis.

To get a sense of Berry's concept of a discipline, it may help to contrast it with common usage. One kind of "discipline" we commonly speak of is technical discipline, a disciplined means of carrying out a task. While such disciplines are important they are not sufficient for Berry since they deal only with means: "people who practice the disciplines of workmanship for their own sake—that is, as specializations—begin a process of degeneracy in those disciplines, for they remove from them the ultimate sense of use or effect by which their vitality and integrity are preserved" (*D&H*: 152). The technical disciplines must be subordinated to "larger disciplines," which deal with what these technical disciplines are for, that is, with the essential problem of living on the land in families and communities. There are three of these disciplines, two of them explicitly concerned with community: "First there is the discipline of principle, the essence of the experience of the historical community. And second there is the discipline of fidelity to the living community, the community of family and neighbors and friends" (*D&H*: 152) (see Figure 7.1). The discipline of principle addresses the issue of continuity with tradition. But Berry does not advocate following tradition blindly. Traditions must always be balanced against the demands of living within actual communities. This is no easy task. "The great moral labor of any age," he notes, "is probably not in the conflict of opposing principles, but in the tension between a living community and those principles that are the distillation of its experience" (*D&H*: 153).

The discipline of community begins within the family; it is there that we first confront the need to harmonize principle with the business of actually living with other people. From there it extends to the local community and thence to politics and government. Berry opens a chapter on "The Ecological Crisis as a Crisis of Character" by quoting from the Confucian four classics: "wanting good government, . . . they first established order in their own families; wanting order in the home, they first disciplined themselves" (*UA*: 16).

Figure 7.1   Berry's Hierarchy of Disciplines

Ends/Meanings	The Larger Disciplines:
	faith—faithful living: dedication to order; acceptance of limits community—loyalty to family, neighbors, land principle—continuity with tradition
Means/Vehicles	Technical Discipline:
	know how

Though necessary, the disciplines of technique and community are not enough. The discipline of faith is necessary also, because "we do not finally know what will be the result of our actions, however correct and excellent they may be. . . . an essential part of a discipline is that relinquishment or abandonment by which we acknowledge and accept its limits" (*D&H*: 156-7). The reason to act responsibly is not because such action guarantees a certain outcome; we can't predict the future that well. Instead, we do so because it is the right thing to do, culturally or morally. Berry makes much of a farmer's comment that he "kept the weeds out of his crops for the same reason that he washed his face." Such an attitude comes not from "technique or technology," nor from formal education, nor even from principle. Instead it is a development in human character that comes from discipline, a passion for excellence and order which grows from the practical acceptance of culture and the thoughtful practice of a way of life (*UA*: 44).

Faith then, is the "ultimate discipline." It need not be faith in a god, or in a feeling of cosmic connection, but it is at least faith "in the propriety of one's disciplines." The dedication to one's way of living that makes it possible for it to be passed on to future generations rests on the conviction that it is worth being passed on. Such a conviction shows itself in the way people live, in the consistency of their actions (*D&H*: 157). Culture helps us maintain a consistency through analogies, metaphors, and rituals. It makes "faithful living" in some broad sense possible. To be consistent we need to understand the ways in which our behavior within the family, for example, is analogous to our behavior in the local community and the natural world.

In part we are unable to keep the larger disciplines of community and faith because we can no longer recognize the kinds of integration we need. One of the effects of the modern world has been to dis-integrate us, he feels; we divide our lives among artificial roles. Who we are at work may be entirely different from who we are at home; our ethics in one setting may be at odds with our ethics in another. This is the theme of "The Body and the Earth," the central chapter of *The Unsettling of America*. At one time, in America as well as the rest of the world, people would have lived in intimate connection and commitment to a certain piece of land. Their commitment to that land would have been inseparable from their commitments to themselves, their spouses, and families, for the land would have been expected to be the home of their descendants. "The farmer's mind, his body, and his farm . . . [were] a single organism" (*GGL:* 143; see also Berry 1984: 28).

In too many cases the demands of industrial agriculture have forced farmers to work too much land, to work it too hard, and to care too little about its future. The result is a "disintegration that is at once cultural and agricultural" (*UA:* 123). In our families we still do recognize (and struggle to uphold) some of the cultural disciplines Berry is concerned with. In marrying and having offspring we do try to make disciplined commitments to those who will come after us: we embrace the technical disciplines of parenthood and spousehood, community disciplines of fidelity, service, and tolerance, and the discipline of faith: we do what is right by our children because it is right and without knowledge of how they will turn out. But as Berry points out, the generations to come will need homes and fertile land as much as we do now. To practice these disciplines in the family but not on the farm, as in much of American agriculture, is both inconsistent and irresponsible. And outside of a system of connections and responsibilities such "special" commitments are likely to seem arbitrary, unnecessary, and to be impossible to sustain.

Culture then, should allow us to recognize how the same balance between principle and living that is manifested in the family might be struck in agriculture, and it does. The archaic term "husbandman," for example, signifies a link between a man's roles as husband to a wife, farmer of a certain piece of land, and steward of the Creation or guarantor of the future. Berry describes the links thus:

> A farmer's relation to his land is the basic and central connection in the relation of humanity to the creation; the agricultural relation *stands for* the larger relation. Similarly, marriage is the basic and central community tie; it begins and stands for the relation we have to family and to the larger circles of human association. And these relationships to the creation and to the human community are in turn basic to, and may stand for, our relationship to God—or to the sustaining mysteries and powers of the creation (*D&H:* 160).

The term "husbandman" is thus an artifact of a responsible culture, now largely gone, in which the proper links between the disciplines of technique, community, and faith were present and recorded in the language. With the dissolution of that culture we now find ourselves puzzled, wondering what analogies between farming and marriage can possibly have given rise to such a term. But our puzzlement reflects our inability to conceive what truly responsible action would be like, Berry maintains.

Laden as they are with Confucian concepts (and ties to Stoicism, Existentialism, Marxism, Anarchism, and Christian distributism) Berry's writings are not easy to categorize or comprehend. Yet this discussion should begin to explain why and how culture and character are so central in Berry's writings. The ailing culture we now have has been so weakened by promises of unlimited technological progress that we are rapidly giving up old disciplines, forgetting the patterns and analogies that have sustained us. Without a viable culture to provide these disciplines or sufficient character to transmit them, we are rapidly becoming dehumanized, for it is culture and character that make us human. However eclectic his sources, in these themes Berry is not exotic or eccentric: the prominence in American politics of issues of character and the centrality of the family in American life make that clear. Ultimately, regenerating culture will be a matter of survival:

> If these connections do necessarily exist, . . . then it is impossible for material order to exist side by side with spiritual disorder, or vice versa, and impossible for one to thrive long at the expense of the other; it is impossible, ultimately, to preserve ourselves apart from our willingness to preserve other creatures; . . . it is impossible to care for each other more or differently than we care for the earth (*UA*: 123).

Berry—Hope

But can we regenerate culture and re-integrate spirit, body, and earth? Does human nature hold out any prospect of hope? Berry and others are convinced that a viable agri-"culture" can be regenerated, not by universities or even humanistic extension programs that might be offered to farmers, but by the great population of dissatisfied people themselves. No matter how attractive the glitter of a consumer culture, there will come a time when its inadequacy as a substitute for meaning, a sense of the larger disciplines, and a sense of rootedness becomes evident.

But how to start this recovery? There are problems both in recognition and action. We can recognize our interconnectedness by recognizing analogies between what may have seemed disparate aspects of our lives, recognizing the implications of the "husbandman" analogy considered earlier. By reflecting on the purpose of marriage, recognizing the importance of its rituals, its commitments, and the cycles it symbolizes, we come to know what good farming might mean. But having recognized, we must act. We restore culture not so much through political, economic, and social institutions whose problems are symptoms of cultural disorder, but by living differently. The problem is primarily personal: "ways of life change only in living" (*UA*: 131). It will not be an easy change, but a "movement . . . uphill" that will require several generations (*UA*: 45). Once enough people have changed, the needed changes will have largely occurred; institutions will then be adapted to match the new way of living.

That we frequently do not see this and continue to locate our problems in social, political, and economic institutions is but another indication of our failure to lead integrated lives. Protests against nuclear power plants, for example, are no more than tantrums if we do nothing to free ourselves from our need for electric power. Under the stimulus of righteousness it becomes

> easy to assume that if only "they" were as clear-eyed, alert, virtuous, and brave as "we" are, our problems would soon be solved. This notion, too, is patently false. In the argument over nuclear power—as in most public arguments—the division between "us" and "them" does not really exist. In our efforts to correct the way things are, we are almost always, almost inevitably, opposing what is wrong with ourselves. If we do not see that, then I think we won't find any of the solutions we are looking for (*GGL*: 164-5).

The solutions will come in what Berry calls a "complete action," an action which one takes on one's own behalf, which is particular and complex, real not symbolic, which one can both accomplish on one's own and take full responsibility for. Growing one's food in a home garden is the exemplar; such an action is also a "solution of pattern" that restores "broken connections" and in which all the consequences are positive; there need be no tradeoffs:

> In gardening, for instance, one works with the body to feed the body. The work, if it is knowledgeable, makes for excellent food. And it makes one hungry. The work thus makes eating both nourishing and joyful, not consumptive, and keeps the eater from getting fat and weak. This is health, wholeness, a source of delight. And such a solution, unlike the typical industrial solution, does not cause new problems (*UA*: 138).

In sum, the context of hope for Berry is in personal acts which are both liberating and satisfying. It is a program which is radical, yet modest in the sense that its viability comes from each of many people taking delight in small changes that bring skill and self-reliance into their lives. Berry's message resonates with home gardeners, those who make jams or can vegetables, those who hunt, fish, and gather, those who build or fix things in their basements and garages. Yet if a critical mass of people change their lives in small ways, great change will result.

Berry—Freedom

Because the cultural transformations must come from within, Berry places great emphasis on the possibility and the meaning of freedom, or more precisely, of human agency. Freedom is not a condition given to us by a government; we free ourselves only by living differently. Berry writes:

> The going assumption seems to be that freedom can be granted only by an institution, that it is the gift of the government to its people. I think it is the other way around. Free men are not set free by their government. Free men have set their government free of themselves; they have made it unnecessary. . . .
>
> A person dependent on somebody else for everything from potatoes to opinions may declare that he is a free man, and his government may issue him a certificate granting him his freedom, but he will not be free (*D&H*: 129-30).

We free ourselves by practicing the larger disciplines within a viable culture.

To some, like Walters, Berry's "culture" and "discipline" will seem nothing more than institutions for conditioning us to behave according to someone else's value system. While it is true that Berry sees culture as a guide, restraining our activities, a true culture will be a product of our own heritage and not a conspiracy foisted upon us. And it will empower as well as guide us. By offering skills and wisdom rather than pat answers, it will put us continually in the position of practicing responsible agency by making judgments about the best way to farm, for example. It will put us also in the position of molders of culture since we will be the ones to regenerate culture in the next generation; our actions, not the products of artists and literati, will shape the culture in which our successors live.

In these ways culture enables us to be free agents, but it also leads us to recognize that absolute autonomy is an illusion. As one recognizes one's agency, one recognizes one's responsibility, and appreciates that certain things must be done because they are necessary or right. The awesome recognition of agency makes it clear that nothing else in the universe will do what needs to be done—not legal institutions, acts of divine justice, economic or ecological pressures—nothing except the action of individual humans. Hence "a person can free himself of a bondage that has been imposed on him only by accepting another bondage that he has chosen. A man who would not be the slave of other men must be the master of himself" (*D&H:* 129). Berry contrasts those who falsely claim to be free with those who have fully accepted responsibility for their lives. The former, finding something about society not to their liking, see the cure as

> "autonomy," another mythical condition, suggesting that the self can be self-determining and independent without regard for any determining circumstance or any of the obvious dependences. . . . There is, in practice, no such thing as autonomy. Practically, there is only a distinction between responsible and irresponsible dependence (*UA:* 111).

Irresponsible dependence is dependence on others or on institutions for what we can and ought to do ourselves. Responsible dependence is an interdependence; it recognizes our real dependence on nature and on human communities and includes moral obligations to others as well as dependence on the soil that sustains us (*D&H:* 156; Campbell 1990).

There are implications of this view of agency for agricultural policy and agricultural research. It suggests that humans, guided by culture, can and will solve the big problems of their existence on the scale of individual lives and small communities so long as they are not disrupted or "unsettled" by outside forces. The role of government should not be to intervene in viable cultures but to protect communities from destablizing interventions. Government sponsored research should supply individuals and communities with tools for provisioning themselves in ways that enhance individual competence and community solidarity and conviviality while maintaining the fruitfulness of the land.

Walters

Walters—The Nature of Humans

While Berry stresses what humans *can* be, with discipline and commitment, Walters focuses on our stupidity, the sheer lack of "mental acuity" of most of humankind. And the context of his concern is justice, not responsibility, a justice that is rare in a world where political, economic, and class conflicts are the inevitable manifestation of human nature. This concept of human nature bears the stamp of a heritage of radical agrarianism: competition between the farm and the institutions of market capitalism is seen as inherent. Walters looks to the nature of the human to explain both why farmers must organize to compete and also why they have historically failed to do so.

Walters accepts that we are lazy and exploitative—"all human history is finally a showdown between the human psychology of acquisition and the possibilities of physics"; "it probably always has been—and always will be—the aim of most human beings to secure perpetual revenue for themselves and their families independent of future effort" (*U*: 54, *AT*: 280). Human nature thus engages us, willingly or not, in an economic struggle against one another. In a highly structured society such as our own, this struggle is among sectors of the economy—farmers, merchants, bankers, and so forth. Economic justice is a condition of balance among these sectors, and in the long run is necessary for human survival. The problem then is not to get rid of the class conflict, which is an unavoidable part of the human condition, but to make sure that people, particularly farmers, can compete effectively and hence maintain an equilibrium.

Walters treats farmers' failure to compete effectively as a failure in rationality, which in turn reflects the weakness of human nature. It is a failure of rationality in two senses: first, it is a failure in comprehension—many farmers were not smart enough "to see what had to be done." But it is also a failure of courage—even those who saw what had to be done did not act rationally; impatience and greed, other components of human nature, subverted rational action.

Thus, where Berry emphasizes faith, Walters emphasizes realistic appraisal of social realities. The following passage from *Holding Action*, his history of the 1968 NFO milk holding action, contrasts the socially aware farmer with the victim of illusion:

> As he dumped his milk, the farmer suffered. . . . He could dump because he saw how his farm fit into the terrible mosaic of the whole. If he were dumping only as an individual, the act could rate as no more than a tantrum, but dumping as part of a farmers' collectivity became meaningful, and this meaning made it possible for him to do something that ran against the grain.
>
> Farmers who still saw themselves as rugged individualists couldn't dump at all. . . . The overview of the rugged individualist ranged no further than the farm, this crop, that cooler full of milk, and all else "out there" was too complicated to grapple with. The individualist could only fall back on God and nature—and prayer—if his stuff wasn't fetching what it was worth. Like the die-hard in any pact with futility, he would

suffocate never realizing he was being choked to death by the marketing system (*HA:* 171).

Here rationality is seen as an understanding of a wider context: the market system. Culture and religion only get in the way of such comprehension. They may do exactly what Berry says, in providing a matrix for meaningful commitments, but in so doing they are serving the interests of others (see Peterson 1990).

But even accurate assessment is not enough. A farmer must protect that assessment from other parts of his nature, particularly from greed. The NFO's great problem in organizing farmers was convincing them to trust their long term rational assessments over the short term rationality of greed. The selfishness of human nature is not at issue here: Walters accepts selfishness as a fact of existence necessary to the workings of society. The problem is incompetence—humans all too often lack the "mental acuity" to work effectively toward the goals human nature sets for them.

Like Berry, Walters sees mass culture and contemporary educational institutions as part of the problem. But while Berry sees these as subverting genuine culture, Walters sees them as imposing a culture that subverts common sense and agrarian consciousness. For him culture is a conspiracy perpetrated by the dominant interests in society. "First, money—lots of it—had to seize control of the intellectual advisors of the community, the journalists, the teachers, the ministers, so that they would sense the needs of the great industrial combines." All this was done secretly by organizations ostensibly on the farmer's side: the "bond between [Farm] Bureau leaders and the powerful classes could not be allowed to reach a level of full consciousness. . . . great pains had to be taken so that no . . . Bureau official would come to feel himself subservient to the ultimate authority" (*AT:* 47). Even the 4-H movement was a means of co-option. It taught individualism over class solidarity:

> business and [Farm] Bureau arrangements slipped money to the 4-H movement. Starting slowly and building momentum, the process taught kids to grow things, rig records, and when the farm boys finally graduated into farming they believed the highest aspiration of all to be besting one's neighbor, not an improved industry (*AT:* 47).

The problem is not just that culture and education are in the control of others, but that they are inherently distorting; they dehumanize their possessors as much as their victims. Those who oppress farmers, for example, are "not abstractions, holding companies, corporations or tax-free institutions" but "flesh and blood people," victims of a feedback process of greed and power that has so unbalanced them that they "see themselves as benefactors of mankind while at the same time they pursue their nihilistic objectives. . . . [and] mix moral insensibility with a vanity tantamount to megalomania" (*AT*: 280, 277).

It all adds up to a pretty dismal prospect for the future of humanity. Institutions are corruptible and frequently corrupt, well-meaning persons become transformed by their own power into tyrants; education, culture, and religion make people into willing victims; abstractions and icons interfere with accurate perception, and the public is left "too dumb and too bewildered to demand an explanation."

Walters—Hope

Yet Walters does see a way to restore mental acuity and redress the prevailing inequality. Since to do this we must see and comprehend with great clarity the dilemma we're in, *raising consciousness*, in a strong sense, becomes critical. Indeed, one of the greatest challenges of social change is to cast off the "brainwashing" of "the powers that be," a formidable undertaking in a society that continually projects distortions (*AT*: 47, 275ff).

Walters finds two sources of hope. One is characteristic of his NFO works, the other of his eco-agricultural writings. The first is the hero, the individual who for no apparent reason transcends his or her background. The exemplar is Oren Lee Staley, NFO president during the holding actions of the late 1960s. "Others might think faster and research deeper and keep their appointments closer, but they couldn't make a single man move. Staley could" (*AT*: 254). Under Staley's leadership farmers suddenly perceived their plight, and were able to act rationally to change it.

To explain how heroes can emerge from among brainwashed and disheartened farmers, Walters draws upon neoplatonic and Christian motifs of enlightenment, grace, and the millennium, quite as much as Berry draws on Old Testament motifs of the specialness of place, the cyclicity of life, and the sacredness of necessary work. Like the ancient Israelites or the early Christians, American farmers are a chosen group, currently undergoing a great period of oppression and trial. And like the prophets who led these people out of their difficulties, ordinary people like Staley—"a prophet on a par with Elijah" (*HA*: 8)—will arise some-how, chosen for their historical missions. In Walters' description of Staley's career we find assembled all the elements of the motif of grace: the common man, in a unique historical situation, discovering his vocation and, through struggle, acquiring wisdom, power, and the burden of acting a major part in a drama of historical inevitability. He is thrust into NFO politics by being made a regional vice-president at an inaugural meeting in 1955. He comes to the ideas of contract pricing and holding actions through a peculiarly agrarian revelation:

> While Staley was riding a cultivator through waist high corn he realized that there could be no solution to low farm income as long as farmers relied on either government payments, or individual marketing. Plain logic suggested that there had to be an overall premise—an overall structure (*HA:* 18).

Then and there he determined to make contract pricing the goal of the NFO. There followed periods of doubt following the failed holding actions in 1962, of bitter struggle for control of the NFO in 1962-3, and of great leadership, in which Staley with a few quiet words calmed an angry mob of NFOers (*HA:* 22-4, 27).

In this view, human frailty is surmounted by a process which, by definition, comes from outside humans or at the least from some peculiar internal process equally inaccessible and unpredictable. It is a good thing that it is inaccessible, for this makes it immune to the brainwashing of the "powers that be"; but because it is unpredictable social changes do not necessarily come when they are needed. On the contrary, history shows that injustice may persist for long periods before a hero comes.

With the development of Walters' eco-agricultural views in the early 1970s, he began to treat the problem of recognizing injustice as a physiological problem with a physiological remedy, rather than an inevitable condition of the majority of humankind. The general idea, which Walters shares with many other alternative agriculturalists, was that a combination of insufficient trace nutrients, nutrients in unnatural soluble states, and insufficient organic matter produced weakling plants that would not have been able to exist under natural conditions, were unable to protect themselves against pests, and were inadequate for human or animal food, though in ways that conventional nutritionists failed to recognize. With publication of the *Acres USA Primer*, Walters had broadened the class of food-induced diseases to include medically orthodox conditions like cancer as well as a "general loss of mental acuity" (*AP:* xii). In issues of *Acres* the perspective was pushed further by Walters and by other contributors. In a July 1982 commentary on natural foods he described the symptoms (and consequences) of the modern malnutrition, the reasons "why most people are sick, ailing, losing their teeth, suffering impaired vision, are tired and dragging their hip pockets—why hospitals are filled to overflowing, why mental diseases now approach epidemic proportions" ("The Last Word," *Acres* 7/82: 38-9). The same cause—"subclinical malnutrition," "debased foods"—was responsible for all the other sorts of "madness [that] stalks the streets": sexual deviance, sadism and torture, violent crimes, juvenile delinquency.

Thus, not only were we putting at risk the fundamental regenerative processes of nature, but also the mental integrity and biological competence of human beings. We are thus in an historically unique situation, able to poison ourselves so insidiously that we are unable to recognize we are being poisoned. The problem is an immediate one, for the longer we allow debased food to dissolve our mental acuity, the more difficult it will be to acquire the mental acuity to recognize the problem and do something about it ("The Last Word," *Acres* 7/82: 38-9).

While biological explanation might seem to place the problem beyond recourse, for Walters it also provides the best basis for a solution: once we have recognized that we are sick and understood the cause of our sickness, we can cure ourselves. While we may be born into a world with a subclinically defective food supply, our plight is not irreversible if we insist on an ecologically sound agriculture. There are advantages then to seeing the problem as material and somatic: we are like machines that need fixing, and we can rest assured that repairs undertaken according to natural laws of nutrition will really work (see Figure 7.2). No longer do we need to wait for a prophet to arrive; we have a real prospect of solving problems of mental defect and injustice if we can bring enough people to recognize that they are among the malnourished.

While Walters' examples may seem extreme, his presentation of the problem is neither original nor incompatible with the broader themes of ecological progressivism. Throughout the centuries many have attributed defects of human character to environmental causes, and particularly to food, and seen the key to a more just and rational social order in a nutritional revolution. In American history Walters' best-known predecessor is Sylvester Graham, who invented the Graham Cracker to combat weaknesses of the flesh. Graham was concerned with questions of why we are not better than we are and give in so easily to conditions harmful to us, concerns remarkably similar to Walters' concern with the lack of mental acuity of modern Americans (Nissenbaum 1980).

Of broader importance here is the idea of technical solutions. A large part of the agricultural community sees a crisis in modern agriculture that is at once social, economic, technical, and personal, and believes that crisis must be dealt with by fixing the systems responsible. It is confident that these systems behave lawfully and that their modes of working can be determined, and hence that they can be fixed. Walters gives more attention to the need to "fix" humans themselves (rather than socioeconomic systems) than do many in this camp, but even on that issue he is not as extreme as he may seem. Walters shares the reductionist orientation of modern behavioral science, and he is perhaps

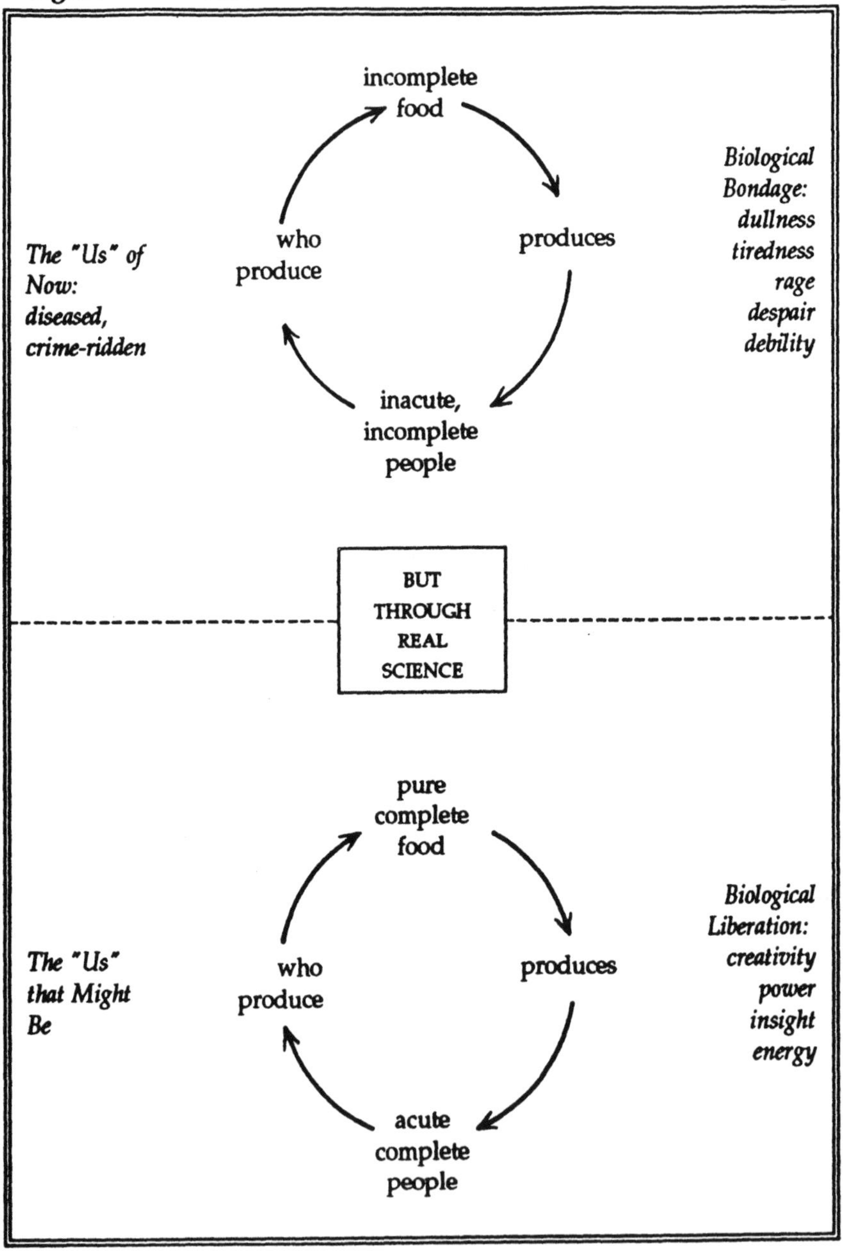

Figure 7.2   Walters: The Dilemma of Malnutrition and Bondage

more in keeping with that spirit in maintaining that the repairs humans need will come in biochemical form—food—rather than in a social or organizational form such as improved interdisciplinary education. His views are, however, far apart from Berry's: while they share a belief that a transformation in persons is necessary, they have very different views of how that transformation can happen.

Walters—Freedom

While Berry's idea of freedom stresses personal responsibility, in Walters' conception freedom is a biological condition that leads to a political one. Without adequate diet we will be unable to take effective action to meet the demands of survival and justice, for the main effect of this "subclinical malnutrition" is to constrain what we might become. It is therefore through true science and the knowledge of nature it provides that we can finally free ourselves. This knowledge, transferred to farmers, could then liberate the rest of us, as well as lead us to an agricultural policy that is both ecologically and economically correct. For Walters then, to be free is to be able to use our minds as they were meant to be used, with acuity of observation and reason.

Walters is less forthcoming about what freedom would be like. *Acres* articles and letters do suggest that we might discover potentials in ourselves far beyond what we currently think available. We would, for example, live much longer since one of the most widespread effects of the current food supply is "degenerative metabolic disease" (*AP*: xii). We would be less tired (Scheresky, *Acres* 8/82: 32). We might be able to exercise our free will in acts which are now impossible for us, such as levitation (P.S. Callahan, "The Real Secret of the Pharaohs," *Acres* 1/82: 10-13). The most important achievement, however, would be the recovery of "mental acuity . . . the only protection one can have in this world, . . . the only security" (*CEA*: 101).

One policy implication of Walters' view is that the route to progress is through research in the sciences of nutrition and fertility. These sciences cannot continue to be practiced as they have been in the past several decades, however: the grants ("bribes") of the petrochemical industry have hitherto distorted the content of these sciences or led to the neglect of the most important research programs. Once corruption in science has been eliminated, biology will be able to progress rapidly, especially if biologists are wise enough to build on foundations laid by past workers, like W.A. Albrecht, or to utilize the work of the few current scientists who have not "sold out" to big oil. Once such science is developed and applied to agriculture we can expect the kind of vibrant democracy Berry envisions. But it is futile to expect the mass of society, acting on its own, to liberate itself from a physiological ailment; instead, it must be liberated by a truly scientific agriculture. Again, Walters' views may seem peculiar and extreme, but his theme that disinterested experts must provide the basis of an ecologically viable society in which people will be independent and prosperous is a common one in the literature of agricultural controversy.

Battelle

Battelle—The Nature of Humans

From the perspective of the Battelle report the question "what are humans like" is apt to seem silly. One might be told to go out and look at them if you want to know what they're like. This attitude, that the problem is either not a real one or not a very interesting one, is a mark of a status quo ideology. Both Berry and Walters are promulgating radical changes and need to ground the possibility of those changes in conceptions of human nature. By contrast the Battelle report neither seeks nor foresees radical changes; one reason they are seen as unlikely is that the current institutions of agriculture are so well suited to human nature. Thus, human nature is in large part taken for granted here and is not, in contrast with the other works, an issue for discussion.

The principal context in which assumptions about human nature are made in the report is one of how we are to achieve progress. Progress in this case means a raised material standard of living, or general prosperity, and to understand how it comes about it is necessary to know some things about the humans who will receive it and who, through their participation in markets and as policymakers, have a key role in making it happen. The achievement of progress, though a product of human activity, is not presented as requiring much purposefulness on our part; instead, the future that will come is seen as an extrapolation of economic forces and technological trends. It is the totality of actions that will constitute these forces and trends; whether they are deeply felt or well-thought-out will make no difference, for such acts will be submerged and neutralized by the acts of others. Still, we may be assured that the result will be progress (Figure 7.3).

The image of human nature that underlies this perception incorporates the human failings that Berry and Walters acknowledge. Humans are lazy, avaricious, and not terribly astute or wise. Unlike Berry and Walters, the Battelle report does not foresee any possibility that they will

Figure 7.3   Human Nature and the Possibilities of Progress

	We need to solve the problem of	Yet humans are	But by means of	We can reach the desired ends of
Berry	meaningless-ness instability ecological disintegration	lazy selfish aggressive easily duped bewildered alienated	Culture	meaning stability health
Walters	injustice ecological stress apathy	selfish lazy stupid incompetent malnourished	Leader-ship and Real Health	justice empower-ment ecological order
Agriculture 2000	inefficiency scarcity	selfish lazy stupid	The Market	greater efficiency prosperity satisfaction

change much. The social machinery must accept and incorporate human weaknesses, and through the institutions of the market and representative democracy, it does. The report concentrates on humans' economic behavior. It treats this as our most important behavior with regard to agriculture, and in general it presumes that our economic actions will be attempts to maximize individual self-interest (though, as we shall see, it is ambiguous on the issues of what self-interest is, whether we are able to recognize it, and whether it would really be a good thing for the economy if we did follow its lead). Although this image of human nature has been abhorrent to many, it is not necessarily either gloomy, immoral, or amoral. In the hands of Adam Smith, for example, the economic rationality of humans was the force that stabilized and optimized society. No one could get away with mischief: consumers and producers, workers and employers, renters and landlords were all smart enough to realize that disingenuous behavior will be counter-productive. Prices would stabilize at a level that was both as low as it could be for consumers—since the goods could not be produced any more cheaply—and as high as it could be for producers, since raising prices would force trade elsewhere. At times *Agriculture 2000* endorses this view: "The U.S. agricultural system today is a prime example of a free enterprise industry" (Battelle 1983a: 169). Here the market is seen as an optimizing system in which each participant behaves rationally.

We get clues to what *Agriculture 2000* thinks humans are like not by knowing that they are self-interested, but by being told where, in the view of the report, their self-interest is likely to lie; i.e., what in a free market they will choose to buy. In its discussion of farm automation, for example, the report gives some indication of what humans are presumed to want: relief from monotonous physical work, for one thing. Thus there will be a new generation of smart tractors and combines that will lessen demand for human labor and human skill: "future grain combines will relieve the operator of most control functions, leaving [the operator with] the responsibility of monitoring the cutting, threshing, separating, and cleaning processes and diagnosing failures" (127). Tractors too will have "automatic guidance" systems that will "relieve the operator from the monotonous job of making small steering corrections, and improve accuracy in high-speed operations" (118, see also 127).

From the Battelle viewpoint, these new machines are progressive in that they are more efficient and easier to use, and because progress is equated with free choice, the report is also assuming that all things being equal, farmers would choose these machines over other types of tractors in a free farm-machinery market. Farmers are seen here as seeking a kind of freedom: a freedom *from*—drudgery, stress, muscular tension, and also a freedom *to*—have leisure, relax. To those with different notions of human nature, like Berry, these same tractors will bring a loss of freedom: operators will no longer have as much control over the machine; there will no longer be an opportunity to grow in experience, to become a master tractor jockey and take pride in one's skill. In Berry's view, if farmers' do buy such machines it will be because the market has ceased to present them with the full range of alternatives or because they have been seduced by a "standard of living . . . entirely material and entirely urban" (*UA* 62). With Battelle as with Berry and Walters, then, particular conceptions of what people are like are continually and intricately woven into the visions of the agricultural future, even when the question of human nature is not explicitly addressed.

Battelle—Hope

The Battelle report makes a point of accepting people as they are. Questions of how we could or should change are moot: human nature being what it is, we are not likely to change, nor do we need to. Finding a basis for hope, then, is less important than for Berry or Walters. Yet here too there is a conception of how human nature is to lead to, or at least to permit the future we hope for. In this view our weaknesses are not obstacles to be overcome but part of the mechanism that will get us to a "realistic" future in which we are accepted as we are.

In *Agriculture 2000* the chief mechanisms of change are economic and the dominant conceptions of human nature focus on peoples' participation in the economy, particularly their performance of the roles of producers and consumers. Yet unlike the classical economists who saw fit to emphasize the extraordinary caution and shrewdness that market participation demanded, the report often utilizes an image of the human as a "chump" whose irrationality, gullibility, and affluence can be exploited in the market for the good of all. It does so not by suggesting that such chumpish behavior is not self-interested, but by drawing on a concept of economic optimality based on the growth of exchange in an affluent society (like our own) rather than on increasing efficiency. In such a context it is possible to indulge what others will see as weaknesses in human nature: getting what we wish for, no matter how frivolous it may seem, is good. Indeed this form of economic organization is especially well suited to our natures: the general good (progress and prosperity) comes because each of us behaves in a way that seems naturally good to us.

The consumer as chump is most evident in the chapter on marketing agricultural products. People forge loyalties to brand names, the report notes, even though "the degree of differentiation within a particular category of branded products may be relatively minor. The important point is that the products are perceived as being slightly different in the eyes of buyers" (64). Here a weakness in the consumer's ability to make accurate assessments is presented as something that society must accept, but also as economically progressive: no matter that consumers are buying on some basis other than quality and quantity; so long as they purchase the branded item at a regular rate they (and society generally) obtain the desirable end of stable prices.

The farmer as chump is evident in a description of the farm supply industry of the future: "aggressive dairy feed suppliers will not merely automatically deliver feed on a certain day of the week. . . . [and] restock the shelf. . . . [but also] automatically deduct payment from the farmer's bank account, thus saving the firm and the farmer time and expense in bill preparation and payment" (72-3). Historically American farmers have struggled repeatedly to free themselves from systems in which suppliers or landlords have controlled their accounts and their cash flow. This passage is written for and from the perspective of the farm supply industry; here farmers are presumed to reject their heritage and relax their vigor to allow greater prosperity in another sector of the economy and perhaps in the economy in general.

It is possible to reconcile both these examples with more narrow concepts of self-interest. The buyers of branded food products may be seeking the social status that comes with conspicuous consumption, or perhaps we really do want to purchase from the processed food industry "a set of satisfactions" which include "nutrition" and "health aspects," "eye appeal," and "packaging," as well as "quality," "convenience," and "taste" (64). The farmer may really find automatic deduction a worthwhile relief from the burden of paperwork. But even if the answer is simply that most people are financially lax and rarely drive as hard a bargain as they might, in *Agriculture 2000* this is not something to be apologized for, nor a condition inconsistent with well-being and prosperity; it is simply something to be recognized, and it is right to incorporate that recognition in economic forecasting in all sorts of ways.

Battelle—Freedom

In *Agriculture 2000* participants in a free market—here American farmers, business people, and consumers—are by definition free. Indeed the concept of a free market epitomizes what freedom means here. What consumers buy can be understood as defining their self-defined self-interest. In participating in a free market, consumers are getting exactly what they want, neither more nor less, and can be trusted to know their interests better than philosophers or economists (Figure 7.4). The choices people make in the market thus reflect the same integrity and wisdom as the choices they make in the voting booth, and are equally legitimate.

The analogy between electoral participation and participation in the market is an important one, for it suggests that those (like Berry and Walters) who are disturbed by the chumpish behavior of the consumer can simply be regarded as people with different values, entitled to their opinions so long as they do not inflict them on others. Berry and Walters might respond that consumers buy on the basis of perceived self-interest but that their perceptions may be illusory, products of ideology (in the pejorative sense) that make what one does appear to be what one must and should do. This clash of perspectives on consumer choice is a widely recurring feature of contemporary tensions between establishmentarian and "utopian" or change-oriented outlooks.

But if the report leaves us in a state of absolute freedom, we will have no basis to make predictions, and no basis to plan policy. The report is, after all, concerned with predicting as accurately as possible the future of

Figure 7.4   The Free Person

Agriculture 2000	I am doing as I please, thank you very much!
Walters	I am struggling to overcome oppression and biological deprivation.
Berry	Aware of my place and responsibilities, I am mastering myself.

agriculture for the next two decades, so that we will be able to plan wisely to secure that future. The more unconditioned the freedom attributed to participants in markets, the more difficult it will be to place confidence in such predictions. We may be able to project with some confidence the size of the population that will need food, but what food they will prefer and utilize, where and under what circumstances it will be grown and manufactured, and how it will be distributed would seem to be open questions whose resolutions would depend on decisions made by people acting freely in the market or in the arenas of policymaking. If these questions were left open, it would be impossible to make precise and informative predictions; the predictions would have less chance of being taken seriously by influential institutions, and hence less chance of being implemented. Since the purpose of such projections is to enable us to plan wisely for the future, indeed to prepare for it, it would be irresponsible to say nothing about how humans will behave.

Many of the projections about human doings in *Agriculture 2000* are based on empirical inferences—extrapolations of trends in consumer behavior, for example—but often they are equally drawn from images of human nature: farmers will shift production from hogs to corn as the prices of these commodities vary because it is rational for them to do so, according to a concept of human nature that stresses rational economic behavior. Such claims or impressions may be drawn from empirical studies of sociologists or economists, but the claim that they will hold true for the future is ultimately a claim that they are grounded in some sort of human nature that does not change. What such prediction does, then, is to anticipate the choices of farmers or consumers. We are told that people will choose, will exercise agency, but we are also told what they will choose. If we go one step further and accept the predictions as something to prepare for, then the people themselves are no longer the ultimate choosers, but the recipients of what has been chosen or "projected" for them. They become units of known needs and desires to be satisfied, but they are not the definers and satisfiers of those needs and desires.

A result of this is that the trends and projections easily become more real and more important than the individual behaviors that constitute them. When people are depicted as behaving in a way that permits the trends to be fulfilled, the trends begin to lose the status of empirical generalizations and begin to take on normative significance as indicators of the right direction to take, and ontological significance as manifestations of the forces that are guiding history. Hence trends (or abstractions like "the economy" which are based on trends) become paramount in determining what happens, even while agency ostensibly remains in the hands of individuals. Berry and Walters are equally insistent about what the future must and will bring, but because their predictions are grounded explicitly in transformed humans, they avoid this paradox.

Thinking in terms of the general behavior of a group rather than in terms of actions or negotiations conducted between individuals is also highly compatible with a treatment that sees the passing agricultural scene from the viewpoint of an administrator of policy or a representative of one of the constituent groups, e.g., farmers, consumers, or processors. It invites us to foresee the outcome of any episode of intergroup interaction by portraying one group, say technologically innovative farmers, as actively carrying out a policy, while others, such as consumers, are seen as the recipients or objects of the policy. One party is the "doer," the others are the "done-to." We may switch perspective from one party to another from time to time, and *Agriculture 2000* does this, but so long as we stick to the idiom of trends and generalizations it is much harder to see several parties in a transaction as simultaneously having equal agency.

The clearest example of this pattern of attributing agency is in the *Agriculture 2000* chapter on "The Quality of Rural Life." The chapter begins by representing quality of life as a function of several factors: "work, family, health, community, neighborhood, and other life experiences" (81). The authors concentrate on community on the grounds that "satisfaction with community is a major element in quality of life, and satisfaction with community services is major in determining community satisfaction" (81). These community services include education, transportation, health care facilities, welfare, and so forth (85). Yet even when they have been provided with services, we still have no warrant to expect that "rural residents" will be pleased with their lives since

> individuals tend to increase their expectations when their environment
> becomes more satisfying. As residents *are provided* with higher levels of
> community services and facilities, they tend to increasingly expect more
> from their communities. Therefore, there is a "rising expectations cost"

associated with community improvement, and complete community satisfaction is never achieved (92, ital ours).

The premise here may well be empirically accurate; but the interpretation, which places agency exclusively with one party (those who "provide" services) is one that treats "rural residents" exclusively as recipients, recipients moreover, who are never fully satisfied with what they receive. The needs for community and for a high quality of life are thus depicted as being like needs for manufactured commodities, not as qualities that people realize through the ways they live together.[1]

The policy implications of this view of human agency are probably more familiar than those we drew from the views of Berry and Walters. Here people are seen not just as units of biological needs, but as making lots of choices. We can, however, generalize as to what they will choose, and hence make highly accurate predictions about how they will behave. What we discover from those generalizations is that they will behave in ways that from Berry's point of view are irresponsible and from Walters' are irrational. But the Battelle perspective sees them as inevitable, as manifestations of human nature as it is, not as we might wish it. It recognizes the obligation for policy to deal with problems that arise from these predicted behaviors, and it sees some social mechanisms, such as the marketplace and the democratic process, that insure that problems will be dealt with quickly. By anticipating the problems that will arise, however, public agricultural policy can ease the stresses on the market and the political system.

Notes

1. Chapters 10 and 11 below takes up analogous issues with regard to the attribution of agency in the achievement of social change.

8

Nature

It is hardly possible to talk about agriculture without questions about "nature" somehow coming up. Critics of modern agriculture claim that it exploits nature or destroys natural systems; its defenders either see themselves as doing no such thing, or see their actions as necessary and proper. But what is this *nature* referred to in these statements? Answering this question is difficult. The range of meanings of "nature" is enormous. Even when we stick to a bit of that range—nature as phenomena external to and independent of the human presence—we run into problems of vagueness and incompatible usage, nor will that definition be acceptable to all with views on the future of agriculture. Yet answering the question is important. It seems unlikely that we will be able to talk usefully about what agriculture should be until we understand the various connotations and implications the term "nature" has.

Images of nature involve answers to the following questions: First, what is "out there"? a system of some kind? or just a pile of matter and a set of forms? Second, what capabilities does it possess, or, alternatively, what constraints does it enforce? If nature has "the final word" on what can happen, to what degree is it also the determiner of what does happen? Finally, what is its normative status. Is nature good or bad or does that question even make sense? Do we look to the "out there" to find a better world, or do we look away from it?

What Nature Might Be

Because the term is so slippery (even in the limited senses used here) it may help to lay out three main conceptions of nature, two of them prominent in Western thought at least since Charles Darwin published

The Origin of Species in 1859, the third a more recent addition, more fully developed in ancient Chinese philosophical traditions, but present also in Western thought at times (Glacken 1967; Ekirch 1973). We label these the Linnaean, Darwinian, and Chinese humanist approaches, and they correspond loosely to the approaches taken by the texts-authors representing our three ideologies (see Figure 8.1).

The first of these is found in a highly developed form in the writings of the Swedish botanist, Linnaeus (1707-1778). Linnaeus' short work on the "Oeconomy of Nature" considered many of the subjects now dealt with by the science of ecology: relations between predators and prey, including the provision of a predator and a source of food for each form of life, and the provision that the easily caught would multiply prolifically so as not to be wiped out, while the most efficient predators would multiply slowly enough that their numbers would not outgrow their food supply. In the view of Linnaeus and those who followed (and anticipated) him these relations were part of the design of creation and indicated the wisdom and beneficence of God.

A similar view of nature is found in much modern writing on ecology, if in a secular form: nature is seen as a whole, a set of ecosystems which within certain limits is stable, bountiful, and predictable. It is usually also seen as a finely tuned system, susceptible to damage or replacement by some less good, less bountiful, less "natural" system. Nature can be understood to have acquired its perfection through a natural selection that rationalizes biological systems far more ruthlessly than any engineer could. Here nature is both rational and normative. Because the best ways of doing things are those nature has worked out, questions about how one "ought" to practice agriculture are answered by looking to nature. These themes are prominent in the agro-ecology movement: it is pointed out that there is accumulated wisdom in nature, in genetic diversity, for example, and that to achieve stability in agriculture we must participate in the complexity of natural systems as much as possible. This conception is the basic framework of Walters' view of nature.

Figure 8.1 Three Views of Nature

Linnaeus	Nature's ways are beautiful and bountiful, they reflect God's design.
Darwin	Beautiful and bountiful is what we call those useful natural things that happen to have survived.
Ancient Chinese Humanism	Nature (heaven-and-earth) reflects how to realize humanity.

A second conception comes from Darwin, who saw the phenomena that Linnaeus had ascribed to beneficent design as simply the set of artifacts that had happened to survive. Easily preyed-upon organisms with too few offspring simply didn't last. Hence while what happened in ecosystems might be lawful, it was in no sense stable or predictable. There was no one "natural" ecosystem, and the boundaries that distinguished ecosystems were in any case constantly being altered by geological processes. While natural forces sometimes enforced equilibrium—as when populations of predators became too large for the available prey—nature did not automatically enforce all equilibria for all time. What stability existed was transient and fragile; the only thing that had been sustained was the earth's ability to support some kind of life, and that sustainability was not so much an illustration of some beneficent system as a generalization from earth history.

In this view, while one might still consider natural systems to have been rationalized in some sense, it was not necessary to imitate them in agriculture. Instead, one could alter environmental parameters in all sorts of ways—by adding water, fertilizer, or pest-destroying agents. The mere fact that a crop grew indicated that its essential ecological requirements were being met and hence that an ecosystem existed. Hence the natural was no longer a guide to what one ought to do. This is essentially the view of nature taken in the Battelle report.

The third conception is more difficult to characterize and was, until recently largely foreign to the western tradition. Nature in this view is neither created nor evolved (in Darwin's sense). It is not the handiwork of a creator, whether beneficent or otherwise, nor the accidental product of selection. Both prior views rely on similar conceptions of time, and hence of change, motion, and history. In the third view time is cyclical rather than linear. The problem of how the world has gotten the way it is does not arise, for processes are characterized less by beginnings and endings, than by the phenomenon of returning. All things return: day to day, year to year, soil to life to soil, birth to birth. Our designation of beginnings or endings is arbitrary; on a cycle every point is a beginning and an end.

Such a view underlay the agrarian humanism of ancient China (Fung 1948: 19-20). Associated with it was a concept of the unity of humans and nature. Nature (or "heaven and earth" in Chinese) unites the internal and external, mind and body, human and nature. Thus what we call "human nature" was not seen as detached from the natural world, but continuous with it. It was partly because of this conception of nature that Confucius was so concerned with transmitting past traditions rather than creating new ones. If nature is cyclical and man united with nature, then innovation threatens to break natural cycles and, far from improving things, leads to degeneration. We cannot improve on nature; the best we can do is follow it. We follow in two senses: we must know enough of the cycles of nature to understand them and accept our place within them; and, if we are wise enough, we can hope to act to maintain them and with them all that is essential in family life, good government, and human character.

Like the Linnaeans, the ancient Chinese found wisdom in nature, but it was not rooted in laws set down by a beneficent Creator. Rather, it was intuitive and analogical and came from long experience in living according to nature. Like the Darwinians, the Chinese did not see nature as created for humans or as necessarily answering to human desires, but there were correct analogies among the several levels of, and connections among, individual practice, family life, government, and heaven-and-earth. Wendell Berry's writings reflect such a view of nature.

In the next three sections we examine how each of the authors deals with the questions of (1) what in a nutshell "nature" is, (2) what, if any, purpose it has, (3) why and how natural events happen, (4) how we ought to interact with nature, and (5) whether and in what ways nature is good.

Walters

Walters—What Nature Is

Walters sees nature as a rationally designed, self-perpetuating system, requiring the involvement of humans. As a "system," nature is presumed to be orderly and predictable: Walters can speak intelligibly of "Nature's way" of doing things. Though nature affects us, we are able to act independently of it and affect its workings. It is a system which benefits human beings by enabling us to survive and was designed to benefit human beings, but we humans do not usually manage it wisely. To manage better we need to know more about how the system works; that knowledge simultaneously will help us improve ourselves—not so much in material goods as in health—and help us avoid inadvertently damaging nature. Currently we are managing it ignorantly and consequently damaging it and ourselves.

Walters—Purpose of Nature

For some the question of the purpose of nature will be senseless. Not only will they insist that nature doesn't have a purpose, but they will be uncomfortable with the question itself. They may grant that there are purposeful human uses of nature that will help us survive and prosper, and hence that we necessarily and legitimately use nature for our own purposes, without going the further step and believing that it is the purpose of nature to provide us with a prosperous existence.

Walters takes this further step. Nature is a Creation and reflects an intelligent plan (*AP*: 408-10). Humans are at the head of the Creation, "at the head of the biotic pyramid" (*AP*: 395). This phrase does not refer to our being top predator in the food chain; it suggests that our well-being is the primary purpose of the creation with nature being the means to that end. In Walters' diagram of the biotic pyramid, the human figure holds a judge's balance as if to indicate that it is our business as humans to make decisions for the rest of the Creation.

Yet even if we accept that the world is to be run by and for people we still have said nothing specific about the purpose of natural things or what their fate should be under our rule. It might be that what is best for humans is to recognize our dependence on the rest of the creation. On this view, agricultural sustainability would come from following or cooperating with what we can understand of the larger order beyond the sphere of human affairs. Following nature would not mean returning to a primordial state, but only fitting human arrangements into an established stability that we disrupt at our peril. On the other hand it might be that managing the world for our purposes requires altering nature substantially, perhaps sacrificing other species to our own purposes. By studying stable ecosystems we learn principles that enable us to invent sustainable agro-ecosystems, even though those systems may well require continual management and may differ significantly from anything that would happen naturally. Each view emulates nature, but in a different sense. In the first view, emulation comes through complementing or completing patterns that nature has begun; it implies following nature. On the second view the emulation is through intelligent intervention based on knowledge of natural law; it implies leading nature. Berry takes the first view, Walters takes the second.

When we learn to understand nature from the correct perspective we will discover how to get it to fulfill its intended purposes, argues Walters. It is not that the workings of nature are opaque to human inquiry, simply that we are too dull to recognize truths that are constantly before us. The weeds in farmers' fields, for example, should be read as messages indicating poor farming:

Every weed is an experienced teacher of very specialized subject matter. And the college of nature has a complete staff of such specialists. It is a pity that the classrooms are not well attended, and that farmer students are allowing their attention to be diverted from the lessons at hand (*AP*: 349).

The "college of nature" teaches the purposes of parts of nature. With regard to weeds and plant diseases, for example:

> Fungi involved with plants during the past 15 million years were designed to take down plants that did not deserve to make seed. Such plants do not live up to the Creator's plan, and so they must go. Insects and fungi are the first steps in taking, say, the wheat plant back into the decomposition cycle. If, indeed, insects went after the strongest, most thriving individuals, then there wouldn't be any strong seed—the worst possible natural selection. As it is insects would rather die of starvation than eat the plants that grow in healthy fields (*AP:* 273).

This passage makes clear the strong sense in which Walters sees nature as a system, and also demonstrates how far he has drifted from orthodox evolutionary theory, even though he continues to use the phrase "natural selection." That term has been transformed from an explanation of a mechanism that increases adaptedness and produces new species to a standard for judging how well organisms are carrying out the Creator's plan. To an orthodox Darwinist differential survival would show only that the survivors had greater resistance to insect depredation, not that insects were the Creator's executioners. The concept of the weak organism would be only relative, it is the *weaker* that are most frequently victims of predators.

The passage also makes clear that the goodness and badness of natural events must be judged from the point of view of human beings. It is to our benefit that nature insures only high quality wheat plants survive; we endanger our survival when we ignore this natural truth. It is then that we have the explosions of diseases, pestilences, and weeds and are forced to resort to futile "crisis medicine": pesticides and herbicides, vaccines, and antibiotics (*AP:* 360).

But the answer is not to let the pests have their way and cull out the incompetent; if we farm properly, according to the Creator's plan, the corn borers and root worms will be out of work, because our crops will be grown on healthy and fully fertile soils. The organisms of destruction will then claim for recycling only what has already served our purposes.

Walters—Why and How

Walters' divergence from Darwin's view of nature reflects his concern with first causes and ultimate origins. He wants to know why things are as they are, and when, where, and how the process began that made them that way. He rejects any sort of sufficient reason explanation: i.e, that things are what they are simply because that is what the interaction of force and matter has led to. For Walters such a view is unacceptable because it sheds no light on the moral or systemic significance of events.

For Walters the answer to the why, where, and when questions is the first cause, the Creator. In the glossary of the *Acres Primer*, he defines both "cause" and "Creator." Cause, "as generally used, . . . means precedence." But since so many related events precede a given event it is futile and confusing to talk about causes in the usual way: if we do, "responsibility fades away, and yet it never disappears entirely, before reaching back to the origin of the world. Thus we arrive at the First Cause (See Creator)" (*AP*: 407). He goes on to define "Creator" as "First Cause, and the author of all causes" (408). He believes we need to look for some kind of transcendent first cause because the statistical laws of thermodynamics would not produce life, much less the astonishingly complicated and miraculously beneficent system that sustains intelligent moral life.

Yet Walters goes further. He recognizes a need not only to explain why things are as they are, but also how the first cause acts to make them that way: somehow the Creator's plan must be transmitted to the Creation. He shares this concern with Ehrenfried Pfeiffer, principal American exponent of the Biodynamic Agriculture of Rudolph Steiner. Extracts from Pfeiffer's writings appear regularly in *Acres*, and in one of these Pfeiffer explains why we must suppose an organizing "etheric" force in addition to gravity, electricity, magnetism, and so forth. Etheric forces are not measurable, yet

> without them there would be no growth, no form, . . . no arrangement or rearrangement of living matter. They have . . . been described . . . as the determining or organizing factor . . . .
>
> According to its proper position in the living system, matter is differential and coordinated. The pattern of form, the model or template—many names have been given to the phenomenon in order to get around the acceptance and acknowledgment of an independent etheric energy—is inherited over generations. Chromosomes and genes are carriers of the fundamental pattern, but not the origin of it (*Acres* 8/82: 11).

One might be tempted to equate etheric force with DNA, but to do so would miss Pfeiffer's point: the etheric force is not the mechanism by which biological order is produced and reproduced, it is the order itself. Pfeiffer goes on to link the etheric force with the Creation:

> As any natural law is an expression of the original creative thought, so is the behavior of the etheric formative force an expression of this same creative thought and intelligence. It is an expression of intelligent action, reason, purpose, determination, metamorphosis, and—above all—of the

functional, harmonious relationship between all the entities of nature and the cosmos. It is the primary force of an order higher than that of the physical energies. The latter are executive; the first is creative (*Acres* 8/82: 11).

The centrality of the concepts of first cause and creation in Walters' writings makes it clear that when we encounter his references to natural cycles or when he personifies nature and speaks of "nature's plan," we need to take these references as more than a convenient shorthand for complicated phenomena. For Berry, "wheels" and "webs" in nature are expedient metaphors for guiding our interaction with it, not literal descriptions of nature itself, which remains beyond comprehension. But Walters must be taken more literally. When he writes of the importance of "how the dusts of the soil are assembled, bolted and screwed together to form the cells of the body and allow them to function" or speaks of enzymes as "on-scene engineers in the cell building business," we would be wrong to think that he is only suiting metaphors to audience (*Acres* 7/82:38-9; *CEA*: 3). Images of manufacture are appropriate to a rational process intended to produce beneficial ends for humans, and that is precisely what the Creation in general, and eco-agriculture specifically, are to do (see Figure 8.2).

Walters—Interaction with Nature

What Walters thinks we need to do with nature, as indicated in the previous chapter, is to make it healthy so we can be healthy. We make nature healthy first by understanding nature's processes intimately in order to make them work better. Wild nature is seen as a well-run factory in which waste products are efficiently recycled and long-term sustainability assured through exclusive use of renewable inputs. Understanding these principles, we can then begin "monkeying around" to optimize production for our needs.

Figure 8.2   Nature and Norms

Walters	We discern how the Creation is meant to be and bring that about (we follow Nature's good).
Agriculture 2000	We shape nature to our ends (nature follows our good).
Berry	We complete and complement nature's patterns (we are part of nature's good).

An *Acres* interview with Don Schriefer, one of the new breed of high-tech farming consultants who Walters hopes will revolutionize American agriculture, is a good example of Walters' perspective. Schriefer's concern is with "first principles" of soil-plant relations: "we view the plant as a factory, and we view farmers as plant managers." Once a soil is thoroughly understood, managers can start "monkeying around . . . [to] get the plant growing." Much greater productivity will be possible: corn yields of 140 bushels/acre, highly satisfactory from most viewpoints, are to Schriefer but proof of poor utilization of light and water. Proper utilization based on a sophisticated understanding of soil-plant relations would allow yields of 300 bushels per acre, he claims (*Acres* 7/82: 24; see also *Acres* 8/82: 24-5, 9/82: 24-7).

Likewise, rather than relying on nature to break down dead organic matter, one can improve nature. The best compost comes not from the set of microbes that occurs naturally nor from use of old compost as starter. Instead, we are to go beyond this "backyard stage" of composting. A scientifically formulated microbial inoculant will do a better job of decomposing than the microbes nature haphazardly supplies (*AP*: 218-22, 375-6).

Hence we interact with nature not by rejecting the activism western culture has practiced since the middle ages, but by making sure that our intervention is guided by the best knowledge we can muster. We are the proper masters and managers of nature in this view, and it is illusory and impractical to pretend otherwise. We can, however, manage well using genuine science, or poorly by allowing shallow and shortsighted economic motives to supplant informed judgment.

Walters—Is Nature Good?

Walters sees nature as good. Indeed he sees greater goodness in nature than do either of the other two writers and relies more on it in presenting a plausible case for human progress. This should not be surprising; so much of Walters' hope is invested in the belief that there is an ideal ecological order that it would be unacceptable to find that this order was ambivalent with respect to human existence. Similarly he locates so much of the failure to realize human potential in soils and foods that it would be unthinkable if there were not soils and foods that could release us from this "physiological bankruptcy." To be sure, the current state of nature is not particularly good, but it does hide a great treasure yet to be unearthed, that will fundamentally transform the human condition.

Battelle

Battelle—What Nature Is

In *Agriculture 2000* nature is the out-there, the totality of nonhuman objects, forces, and possibilities which we are in the process of rearranging for our own purposes. Although the whatever-is-out-there of nature is presumed to behave lawfully, and hence to constitute a system in the sense thermodynamicists might use that term, it is not the sort of system Walters conceives, not a wisely designed, bountiful, and intricate mechanism meant to sustain humanity. As well as being shaped by physical forces like gravity and electromagnetism, nature is being rearranged by economic forces. The laws of economics are conceived as true laws of nature, no different from physical laws, except in their sphere of operation. There is no transcendence or special goodness in this vision of nature—nature just is. It has no special place in it for humankind, though as parts and products of nature we are entitled to make of it what we will. Because the authors of *Agriculture 2000* do not see nature as some integrated whole, it is somewhat inappropriate to use the term "nature" to characterize their views. In particular, questions about the purpose and goodness of nature presuppose that there really is a coherent whole that can be the subject of such attributes, even if we do not or cannot fully comprehend it. But if the term is only a generic noun for collecting a great many disparate entities we haven't a better word for, such questions are strained at best. For sake of symmetry we have asked them nevertheless, and the answers have proved surprisingly illuminating.

Battelle—Purpose of Nature

Many would characterize the vision of nature in *Agriculture 2000* as Baconian. In it the world is a collection of raw materials, each with its own interesting properties. If we are clever we can put these together in remarkable and useful ways. Yet in contrast with Bacon the text doesn't present nature as manifesting a vast wisdom or a unique systemic integrity, and since nature doesn't have these characteristics the text takes it for granted that it is acceptable and desirable to reorder nature in what seems a productive way. The moral center of gravity is humankind, not some ideal ecosystem. Hence the text finds no significant moral (nor for that matter, practical) problems in exploiting "underexploited" land (Battelle 1983a: 105), in "direct manipulation of an organism's genetic material" with recombinant DNA techniques (108, 141), or in "synchronization" of estrus in sows (160), techniques which from other perspectives have been considered dangerous, immoral, or both. That parts of nature might not be for human use or that there might be moral constraints on the use of nonhuman entities is never entertained, and institutionalized science is invariably presented as providing the means for using nature.

Battelle—Why and How

In *Agriculture 2000* the issue of cause is dealt with as an aspect of economics. What will happen in our agriculture is what is economic; indeed the push toward greater economic efficiency is almost as inexorable as increasing disorder in thermodynamics or increasing adaptedness in biology. We could even translate the terms of economics *Agriculture 2000* uses into terms of biology—increases in "productivity" (122) or "efficiency" (121) are increases in adaptedness, "progress" is adaptation, "pressure" for change (121, 130) is equated with the "conditions of existence" that produce natural selection. In short, economics supplies the order, norms, and system that nature supplies for Walters.

There is a long tradition of analogy between economics and biology, and particularly ecology. Some early ecologists, like Linnaeus, saw their study as the "economy of nature," and indeed, were studying relations of "production, distribution, and consumption of . . . material needs."[1] For Linnaeus (or Walters) the economy of nature was a description of static relations, not of forces that produced those relations. In the laissez-faire economics of Darwin's time on the other hand, the economy was a product of the combined forces of the self-interests of each participant. Yet even after it ceased to be seen in a Linnaean sense as an unchanging system of "stations" (ecological niches) and relations between stations, the "economy" retained a certain rationality, wisdom, and beneficence, since it was felt that the pursuit of self-interest would best serve the well being of the entire society.

In *Agriculture 2000* "nature" and "economics" are no longer distinct categories: each refers to an ordered collection of units whose order comes from the working of forces. The particular order that exists at a given time is seen as the one that must exist because it is the determined ordering. Further, each term embodies an idea of progress, an assumption that the existing ordering is good but that continued operation of the forces will ensure that future orderings will be better. Indeed, because nature is understood as not having a unified, inherent character, and because humans can and do shape and define it, "economics" in many respects comes to supplant "nature."

Seeing problems in nature as economic problems contributes to their solution, for if problems are essentially economic, the conditions for their solution have exclusively to do with constraints imposed by the economic environment. When incentives to control soil erosion are strong enough it will be controlled; as resources become scarce they will either be used more sparingly or alternatives will be found.[2] We find the means— develop the technologies—when it is necessary for us to do so to survive; necessity truly is the mother of invention. In a sense then, all our problems are already solved—potential solutions already exist. They await only the catalyst of necessity and the money and effort necessity brings with it to become actual solutions. Imminent disaster is not the only source of these incentives; incentives are also provided by the constant pressure toward greater productivity and efficiency.

Good illustrations of this perspective are found in the section on "Developing Technologies." There it is pointed out that "evaporation suppressants applied to seed beds . . . will be used as a means of preventing water loss" (51); that "tissue culture will be used to quickly develop disease resistant varieties that would take years to produce via conventional plant breeding"; that "genetic engineering and plant breeding also will allow the production of plants such as oilseeds with a better overall amino acid balance in the seed protein" (52). What all these statements presume is that the technologies in question will work, and will be economical to employ.[3] The repeated use of "will" makes clear that the problems are in some sense solved; they exist no longer as problems, but only as economic situations which have not yet reached a condition where these solutions are triggered.

The point we wish to raise here is not whether or not this future will happen but the identification of constraints as economic. Both Walters and Berry by contrast will argue that nature really imposes constraints; there are problems technology will never solve, regardless of time, money, and effort. The advantage of transferring causation from the domain of nature to that of economics is clear, however. If the con- straints on what will happen are ultimately economic, then we are the masters of our own fate; what happens is determined by incentives, and incentives come about when threats to our survival arise.[4]

Battelle—Interaction with Nature

We interact with nature by using it for our purposes. As suggested above, economics is the final arbiter of what becomes of nature's raw materials. Hence, rather than taking the view that what is economic is a subset of what is natural, the authors are in effect suggesting that economics defines what is naturally possible and converts it to actuality as well: we discover what is possible by doing it. Some may argue that there is a hidden step here; that we must (and do) recognize natural limits and perhaps problems from which we cannot extricate ourselves.

The Battelle authors do not, for example, promise us plants that need no nitrogen or animals that need no feed, yet their visions of how some serious environmental and agricultural problems will be solved do suggest that we are still far from the limits of possibility—things will be done that may now seem profoundly unnatural.

An example of how economic determinants can displace natural constraints is the discussion of "salt-tolerant crops" for regions where irrigation has led to such heavy concentrations of salts that crops cannot be grown successfully. To solve this problem plant breeders are "screening" varieties for salt tolerance. While the authors note that progress has been made, "at this point, no commercial cultivars have been developed using this technique; however, with time, salt-tolerant germ plasm will be incorporated into commercial cultivars" (105). They go on: "The development of salt-tolerant crops will have a great impact on agricultural production, particularly in the western United States"; and a paragraph later: "As salt-tolerant cultivars of crops are developed, many sectors of the agricultural production system will be affected" (105).

Very likely such consequences will follow if such crops are developed, but the discussion contains no "if." The following sort of reasoning is apparently at work: As salinization becomes increasingly serious, there will be steadily greater pressures to devote more and more funds to finding a solution. Because the economic problem has become so big, a solution will be found for it, perhaps through a more powerful trans-formative technology than genetic screening. The contrast between the view of problem-solving in *Agriculture 2000*, in which constraints are defined in terms of time, effort, and money, and those of Berry or Walters, where they are defined in terms of culture, nature, and the human condition, is striking indeed.

A stronger example concerns availability of irrigation water: "increased withdrawal of ground water can be compared to mining when withdrawal exceeds the natural recharge capacity," the authors observe. The mining comparison refers to overuse of renewable resources: removing the contents of a reservoir more rapidly than they are replaced is effectively to empty the reservoir. Most critical is the Panhandle area where

> water is pumped from underground aquifers which are extremely slow to recharge. No real effort has been made in recent years to reverse the trend of increased irrigation. However, many water conserving irrigation technologies are being developed, as noted later in this book (9).

They do discuss new methods of irrigation, such as drip irrigation, which use much less water—"savings of up to 50 percent compared to conventional systems have been achieved in orchards" (98). Yet the water problem is essentially a problem of distribution and conservation:

> Despite governmental action to avert water shortages, distribution problems still exist, so it is anticipated that farmers will plant different crops and use new irrigation techniques. . . .
>   Concern about vegetable crop production in areas where water is scarce will result in more use of efficient irrigation techniques (96).

Thus as water becomes more scarce, it becomes more expensive and farmers switch to techniques that use less. Potentially these have always existed as part of the realm of the possible, yet not until now have they become economically feasible, and hence actual. While the authors of *Agriculture 2000* recognize that "advanced irrigation technologies will not solve all water supply problems, and are not even a substitute for other systems in all cases," (99) they do not state any limit to this adaptation. Water may get scarcer and scarcer and increasingly more expensive, but there will always be some technique or new crop that can use less of it.

The passage is intriguing because it captures in a nutshell much of what is different about the way those in different camps see progress and possibility. For Berry and Walters it might seem obvious that to withdraw water from an aquifer more quickly than it is recharged will, by definition, be unsustainable. More careful use of water—conservation—would make no difference in the long run as long as withdrawal exceeded recharge. But from a viewpoint in which economics dictates what is possible, and which has supreme confidence in the adaptability of technology to resource constraints, the problem simply looks different. It may be true that currently we are depleting a reservoir faster than it is recharged, but it would be wrong to presume that present economic conditions, technologies, or scientific knowledge are the proper standards of assessment. We need not assume that any particular solution now on the drawing boards will be the solution to the question but only that forces of supply and demand acting on human talent and the raw material of nature will carve out some solution. In a sense this thinking does give rise to a sustainable agriculture. It recognizes enormous variability in supply, demand, and price, as well as in our estimation of what is technically possible, but because the market itself must continue, something will be sustained.

Battelle—Is Nature Good?

Is nature as envisioned by *Agriculture 2000* good? As noted earlier, if by "nature" we mean some transcendent, coherent entity imposing its laws on us, the question cannot be satisfactorily answered—no such nature is present in the report. To be sure we can make a good result—in surviving and prospering—out of natural raw materials, and perhaps in some weak sense this implies a goodness in nature manifest in our ability to do so. But for comparison with Walters and Berry, we can say at best that nature is neutral with respect to the human enterprise. It will not guarantee our welfare nor wilfully destroy us, simply because in the context of those sorts of questions "nature" does not exist.

Berry

Berry—What Nature Is

For both Battelle and Walters nature is mainly a deliverer of goods. It is outside us, we get things from it. Berry replaces the emphasis on things with an emphasis on processes, processes moreover in which we participate and to which we are subjected (see Figure 8.3). These are the cyclical processes of "birth, growth, maturity, death, and decay" (*UA:* 82). These processes are quite as much "in here" as they are "out there." Eliminating the distinction (and opposition) between the natural and the human does not mean throwing out distinctions between self and non-self or observer and observed. Nor does it require us to think of ourselves simply as walking, talking collections of atoms. Nor does it mean indulging in what we may think is our naturalness (*HE:* 147). It does suggest, however, that it makes no sense to think that in doing things to nature we are not also doing things to ourselves.

Figure 8.3   What Is Nature?

Agriculture 2000	Nature is whatever is out there and whatever happens; what is, is natural.
Walters	Nature is what the Creator designed.
Berry	Nature is the larger cyclical process in which we have our being.

There is an orderliness in natural processes—a simple repetitiveness that perhaps implies some notion of nature as a transcendent entity in some weak sense. But unlike Walters' vision of natural order, which holds out for us the prospect of wealth, health, and perfection if only we discover and employ nature's ways, Berry's nature does not offer us an existence very much different from the world we currently inhabit. Following nature doesn't lead us anywhere new, since nature is cyclical. At best we can keep the cycles going, at worst we can disrupt them irreparably.

For Berry then, the main problem posed by nature is not to transform it, but to accept its limits and live within them, and even to appreciate and celebrate those limits as the source of the larger meanings of human life. This is no easy matter. It is easy enough at the level of sermons to dwell on the endlessly mysterious conversion of all means to the ends of sustainability. Berry does this:

> The soil is the great connector of lives, the source and destination of all. It is the healer and restorer and resurrector, by which disease passes into health, age into youth, death into life. . . .
> It is alive itself. It is a grave, too, of course. . . . full of dead animals and plants, bodies that have passed through other bodies. . . . But no matter how finely the dead are broken down, or how many times they are eaten, they yet give into other life. . . . Within this powerful economy, it seems that death occurs only for the good of life. And having followed the cycle around, we see that we have not only a description of the fundamental biological process, but also a metaphor of great beauty and power (*UA*: 86).

But with respect to the life of an individual all this is likely to seem profoundly unsatisfying. The problem is not one of expounding grand truths, but of reconciling them with day-to-day living in the practice of the larger disciplines of community and faith. It is here that culture becomes essential.

Berry—Purpose of Nature

For both Walters and the Battelle report, who see us as standing outside nature, questions about nature's purpose are important in determining the relations between its purposes, if any, and our own: can the two, presumably distinct, be reconciled? They offer two options: the quasi-theological view of Walters, who finds nature full of purposes because God created it to serve a divine purpose; and the wholly secular view of Battelle, which finds no room for purpose in nature beyond what humans impose on it. In both discussions to talk about purpose presumes intelligence, either of a creator or ourselves. For Berry, who sees us as within nature, we cannot talk usefully about the purpose of nature until we abandon commonplace notions of purpose and regard our own existence and purposes as explicable only within the patterns nature displays. For Walters and the Battelle report we understand nature by ascribing to it categories of human purposiveness, for Berry we understand ourselves by placing ourselves within the seeming purpose-lessness (or ill comprehended order) of nature (see Figure 8.4).

If nature can be said to have a purpose in Berry's writings, that purpose is to return. The focus of Berry's writing about nature is cyclicity, a tendency to return "again and again" to the same state, rather than to move along a "road" (*UA*: 89; *D&H*: 139ff). From this perspective most of our attempts to attribute purposes to natural systems or objects have been attempts to find in nature some warrant for our own greed. But while we may have purposes, we fool ourselves if we claim that they are equally nature's purposes.

Understanding this idea of return and understanding that it is not necessarily our purpose leads, in Berry's view, to restraint not ambition, to a willingness to work within the order of nature rather than to seek some end that lies beyond it. He recognizes that we live only by expropriating things (and creatures) from nature, but stresses that we are only "worthy of the world's goods . . . by using them carefully, preserving them, relinquishing them in good order when . . . [we have] had their use" (*D&H*: 113).

Figure 8.4 Purpose, System, and Order in Nature

	Walters	*Agriculture 2000*	*Berry*
Has Nature a Purpose?	yes (strong sense)	no	the question is nonsensical
Is Nature a System?	yes (strong sense): benevolent and rational	yes (weak sense): rational	yes, de facto: system as mystery
Is Nature a Source of Order?	yes (strong sense)	no: order is what humans impose	yes (weak sense): acceptance of our place and limits)

Thus Berry does find a purpose in nature that is beyond human purpose, is worthy of being followed, and that we violate at our peril. But it is not a purpose we can follow to some promised land. Following it leads nowhere but back to ourselves, to our limits as humans, within which we must live and die if we are to do either with dignity.

Berry—Why and How

For both Walters and Battelle knowing natural causes is a means for humanity to achieve its goals. For Walters nature's causal laws teach and enforce correct farming: when we farm unecologically, e.g., in growing a "weakling" wheat crop, nature punishes us with weeds and pests. When we farm as we ought, nature rewards us. In the Battelle report causation takes place through an economic mechanism—what we demand happens if the demand is strong enough. Causation has a similar normative context, for it is through the rational, causal mechanisms of economics that we supply ourselves with goods. For both Walters and Battelle then, what nature does is somehow just. Whatever happens has happened for reasons that given a little perspicacity we can usually discover. If we didn't want the consequences we got, we ought not to have taken the actions that caused them. Both envision a carrot-and-stick universe that guides, even determines our actions and indicates right, wrong, and necessity.

For Berry such guidance must come from culture. Even if nature is in some sense lawful, we still cannot treat it as a trustworthy test that will immediately deliver a just set of consequences for our actions. In the long term if we farm responsibly we will have a sustainable agriculture. Yet we still have no warrant to expect a guarantee that fate will treat us justly; instead, responsible action must rest on a commitment of faith, since "we do not finally know what will be the result of our actions, however correct and excellent they may be" (*D&H*: 156). As for the short term, we act responsibly because it is right, not because we know we will be rewarded by a good outcome.

Thus Berry has little patience with Walters' claim that sophisticated fertility assessments and soil treatments will resolve problems of pest and weed infestation. And even if they did, there are still the random insults of weather and markets. Farming is not fair, "the forest is always waiting to overrun the fields," and the insects to eat the crops. Science is no help. The unique events that are so significant at the level of individual lives and individual farms will never succumb to any precise causal calculus. We can predict only the probability of hailstorms or tornadoes; we must learn to live with, not overcome, the randomness of such caused events. It is best to accept the mystery of natural causation as not to be solved but celebrated, and to realize that we survive despite it and because of it. It is this perspective that leads Berry to stress ritual, religion, and culture in agriculture.

Berry—Interaction with Nature

Clearly we can't avoid interacting with nature and on nature's terms. In Berry's view both the Walters' position and the Battelle position seem to insist on a future quite different from the present, in which we will be more powerful, live longer, and achieve whatever we are seeking (*D&H:* 148). To Berry such a vision is probably unrealistic and certainly unwise: "Much as we long for infinities of power and duration, we have no evidence that these lie within our reach, much less within our responsibility. It is more likely that we will have either to live within our limits, within the human definition, or not live at all" (*UA:* 94). He writes of Chinese landscape paintings populated yet not dominated by humans. The juxtaposition of "towering mountains" and the "tiny human figure" represents "a world in which humans belong, but which does not belong to humans in any tidy economic sense; the Creation provides a place for humans, but it is greater than humanity and within it even great men are small" (*UA:* 98).

Berry's idea of how we are to relate to nature is epitomized in his concept of "kindly use."

> Kindly use is a concept that of necessity broadens, becoming more complex and diverse, as it approaches action. The land is too various in its kinds, climates, conditions, declivities, aspects, and histories to conform to any generalized understanding or to prosper under generalized treatment. The use of land cannot be both general and kindly. . . . Kindly use depends upon intimate knowledge, the most sensitive responsiveness and responsibility" (*UA:* 31).

A scientific approach, leading to the control of nature, does not help much here. If we are to use nature respectfully and relinquish it in good order, then we must know it intimately in its particularity and respond to it with great sensitivity.

Kindly use clearly involves careful work, and Berry has much to say about work, especially farm work. The dis-integrating character of modern civilization has led to the denigration (literally, the "niggering") of work, he maintains; we have seen "the growth of the idea that work is beneath human dignity, particularly any form of hand work." As a consequence work itself has changed: "We have made it our overriding ambition to escape work, and as a consequence have debased work until it is only fit to escape from" (*UA*: 12; *D&H*: 116). Work becomes meaningful when it is necessary, and it is not drudgery if it involves the mind. When one is directly involved with the quality of the biological processes going on in every corner of a farm, one is doing necessary work, one is doing it carefully and well, one's mind is constantly engaged, and one is participating in the processes of nature.

Whether work humanizes or dehumanizes will depend in part on the technologies through which we interact with nature. If the "industrial world" gets between us and the sources of our lives, it is getting between "the mind and its world." When we let machines do our living for us we "diminish life": "we cannot live in machines. We can only live in the world, in life. To live, our contact with the sources of life must remain direct: we must eat, drink, breathe, move, mate, etc" (*UA*: 92). He complains that the current breed of closed-in, controlled-environment tractors are like coffins for farmers "who not only hate farming (that is, any physical contact with the ground or the weather or the crops), but also hate tractors, from the 'dust', 'fumes', 'noise, vibration, and jolts' of which they wish to be protected by an 'earth space capsule'." Those who found such features attractive were those who wished to be dead, "who have found nothing of interest here on earth, nothing to do, and are impatient to be shed of earthly concerns" (*GGL*: 185). It is not that such people are missing the fun inside their artificial environments. Quite the contrary: the things that such tractors protect a farmer from—noise, weather, dirt—may truly be unpleasant. But to try to avoid them signifies both that one has assumed life can be lived outside the bounds of natural responsibility and that it will be worthwhile to live it that way. Berry, by contrast,

> would like a fair crack at living. . . . I acknowledge that the world, the weather, and the life cycle have caused me no end of trouble, and yet I look forward to putting in another forty or so years with them because they have also given me no end of pleasure and instruction (*GGL*: 186).

When people reject such convenience and control they are making "their lives more risky and difficult than before" (*GGL*: 180). Yet they do so because satisfaction "comes from contact with the materials and lives of this world, from the mutual dependence of creatures on one another, from fellow feeling" (*GGL*: 180-1).

There is a way of mixed farming that embodies these concerns. Currently, it is most conspicuously practiced on Amish farms and in Amish lives, though it is not historically or uniquely Amish. Mixed farms will not escape nature; in Berry's view, each should include a woodlot. The trees will be on the farm because they are useful, but also because they represent and sustain the connection between the temporary, fabricated order of the farm and the eternal and wild processes on which farming depends.

An enduring agriculture must never cease to consider and respect and preserve wildness. The farm can exist only within the wilderness of mystery and natural force. And if the farm is to last and remain in health, the wilderness must survive within the farm (*UA:* 130-1; see *HE:* 137-51).

Interaction with nature is for Berry a good deal more central than it is for the other authors—for him the key to sustainable agriculture, social progress, stable rural communities, and satisfaction with one's own life comes through recognition of the natural order and participation within it.

Berry—Is Nature Good?

When we ask whether nature is good in Berry's vision of the world we run into the same trouble as when we asked about purpose. Immediately we must ask "good for what ends?" In the sections on Walters and the Battelle report we took this question to refer to the goodness of nature in providing the means to meet ends that individuals and societies define: longer life spans, better health, economic growth, better control of natural processes. If we stick to this way of posing the question we find in Berry an eloquent ambivalence (see Figure 8.5). Sometimes we can get nature to deliver the goods we want, and sometimes we are just lucky recipients of nature's bounty, but nature fails to oblige us at least as often as it obliges us, and after we have kept the books over a long period we are likely to find only that nature is indifferent to our existence.

Figure 8.5 Is Nature Good?

	Is Nature Good?	Why?	Focuses On
Agriculture 2000	no	Nature just is; we can use it for our own good.	humans
Berry	yes and no	Accepting our place in nature is good for us, even if nature is not good to us.	persons in nature
Walters	yes	Nature is good; it was made for us.	nature

But in terms of Berry's writings it is more appropriate to take the question differently. Instead of measuring nature's goodness on a scale of how well it fulfills human desires, we would do better to define the processes of nature as the exemplar of good. Much of Berry's writing is about learning to be responsive to norms, orders, or powers found in nature. It is by following the disciplines and norms of restraint, necessary work, and return that we find personal satisfaction and sustainability. Hence nature is not so much good *to* us in the Santa-Claus sense, as good *for* us. One of Berry's principal complaints about the images of progress in modern society is that they keep us in perpetual dissatisfaction: whatever we have is not enough. By contrast, a culture based on responsible participation in nature satisfies us with "enough" (*GGL*: 169).

Notes

1. There are also etymological links. Both terms come from the Greek "oikos," meaning house or household. Ecology refers to the science of the household, economics to the management of the household.

2. This argument derives from the conflation of economics and nature; it involves substituting a Lamarckian version of how nature works for a Darwinian version. In Darwin's view organisms that do not adapt to a changing environment become extinct; in Lamarck's organisms adapt so as not to become extinct.

3. Only occasionally is this assumption explicit, e.g., "More farmers will use antitranspirants as a means of inhibiting the loss of water vapor from leaves. . . This assumes that more effective antitranspirants will be developed that do not affect carbon dioxide exchange" (51).

4. There are a few places in *Agriculture 2000* where the authors do raise the possibility that nature may force a tradeoff. With respect to development of corn (or of corn-microbe symbioses) able to fix atmospheric nitrogen, they note that "there is also some feeling that nitrogen fixation in corn can be achieved only with decrease in yields of perhaps up to 30 percent" (53). Here nature is evidently forcing a tradeoff between production and nitrogen fixation in the view of some scientists.

9

Knowledge

Along with images of nature, images of knowledge are of great importance in understanding various visions of agriculture. So much seems to depend on knowledge—on knowing what demands will be placed on agricultural systems, what resources we will be able to count on, how natural and social systems will respond to changes we make. On several issues about knowledge our authors differ significantly: (1) they differ with respect to what knowledge is; (2) while all are critical of knowledge producing and disseminating institutions like universities, they differ greatly in diagnoses of the problem and prescriptions for repair; (3) they differ about who knows best, particularly what farmers know in contrast with scientists or managers; (4) they differ about how new knowledge is to be had and even whether it is available, necessary, or a good idea; (5) they differ with respect to the moral status of knowing; there is disagreement for example about whether new knowledge engenders moral problems or solves them.

Berry

Berry—What Is Knowledge?

For many, and perhaps most of us, knowledge is something we possess more or less of, a commodity. We know how to roller skate or speak french with a certain proficiency, or that easterly winds are associated with warm fronts and oxidation occurs only in the upper level of the soil. We can distinguish when we are acquiring knowledge from when we are utilizing it or simply doing other things.

For Wendell Berry knowledge is not a possession but a way of being: knowing is knowing how to act. The state of our knowledge is manifested in what we do and the circumstances under which we do it, and not by how many pieces of information or ability our action appears to entail. When and how we decide that it is right to call upon particular pieces of knowledge is part of knowing. Knowledge of good farming cannot be extracted from actions and reduced to a manual. To think that it could would be to forget the necessary integration of mind and body, spirit and land. To try to do that would be to try to separate and elevate our cognitive faculties in a way that would destroy the knowledge itself. Berry writes that "there is an inescapable kinship between farming and art, for farming depends as much on character, devotion, imagination, and the sense of structure, as on knowledge [in the sense of possessed knowledge]. It is a practical art" (*UA*: 87). Elsewhere he writes that solving the problems of agriculture requires "accurate memory, observation, insight, imagination, inventiveness, reverence, devotion, fidelity, restraint" (*GGL*: 145).

There are good reasons why it seems both so difficult, and in Berry's view so unwise, to attempt to dissect knowledge out of the matrix of skills, attitudes, and ways of being which make responsible actions possible. The main reason is that successful and sustainable farming in the past depended on all these capacities. One had to have certain skills—knowledge in the usual sense—but one also had to discipline the application of those skills with reverence, fidelity, and devotion, and one had to have the imagination and sharp-wittedness to adapt to new situations. There is no adequate formula for farming in which all of these are terms, for the demands on each farmer are unique and change so unpredictably that farming by formula is impossible. But a healthy farm culture embodies all these attributes, and as one grows up and into a culture, he or she develops them through exercising them. Berry's reclamation of an eroded hillside on his own farm was an act of devotion and fidelity, one which required and strengthened imagination and skill.

All this contrasts in many ways with views which conceive knowledge to be separable from action and yet see knowledge as essential for and even sufficient to guarantee wise actions. Berry disagrees:

> that specialization has vastly increased our knowledge . . . cannot be disputed. But I think that one might reasonably dispute the underlying assumption that knowledge per se, undisciplined knowledge, is good. For while specialization has increased knowledge, it has fragmented it. And this fragmentation of knowledge has been accompanied by a fragmentation of discipline (*D&H*: 95).

For Berry, this problem is exemplified in the goal of modeling or forecasting the behavior of various systems, particularly agro-economic systems, the kind of enterprise that *Agriculture 2000* represents. Such models represent what is wrong with knowledge by itself. Models assume boundaries, boundaries dictated by the specialty of the expert making the model and the purpose and audience of the model. Within those boundaries all is in control because the specialist has drawn the boundaries to include only those factors capable of being controlled:

> having chosen the possibility of total control within a small and highly simplified enclosure, he [the specialist] simply abandons the rest, leaves it totally *out* of control; that is, he forsakes or even repudiates the complex, partly mysterious patterns of interdependence and cooperation, controllable only within limits (*UA:* 71).

As we recall from Chapter 8, a great deal of Berry's concern was directed not at taming wilderness but accepting it, and finding ways of dealing with the wildness that continually enters our lives.

Where Berry will emphasize the moral dilemmas that arise when we act on the basis of what purports to be knowledge, models of detached knowledge, in which knowledge is treated as objective, descriptive, and value neutral, obscure those moral dilemmas. Models which claim to show how the economy works, for example, do not for Berry warrant the claim that those workings are right or necessary. And if those workings are neither right nor necessary, there is no reason to think we really *know* them, since morally mature humans may do something quite different than what the models predict. Berry writes that "we have made our false economy a false god, and it has made blasphemy of the truth [i.e., of true knowledge]. So I have met the economy in the road, and am expected to yield the right of way. But I will not get over. My reason is that I am a man, and have a better right to the ground than the economy" (*D&H:* 121). Here "the economy," i.e., what is claimed as knowledge of economic reality, is presented as disembodied, abstract, amoral knowledge, which in Berry's view is not knowledge at all.

Berry—Institutions of Knowledge

Berry frequently makes the point that many problems of agriculture lie in the overspecialized organization of land grant universities. The disciplines of study are no longer "mutually sustaining and enriching," with each focusing from a different angle on the human condition. As an applied science, agriculture has been severed from the rest of the university and connections between farming and the rest of living have been lost—"farming shall be the responsibility only of the college of agriculture, . . . morality shall be taken care of by the philosophy department, reading by the English department, and so on" (*UA*: 43; also *HE*: 76-97).

Even agricultural science itself has become so highly subdivided that problems are formulated and answered without regard to the total context in which farmers work. In "The International Hill Land Symposium in West Virginia," a 1976 essay, Berry complains of "technologies and techniques" being presented "without [anyone] ever mentioning their economic effects on the farmers who used them—much less their political, social, and cultural effects."

> We would be shown a slide of an improved hillside on a farm, say, in Ohio. We would be told the composition of the soil, the techniques of renovation, the forage plants used. But we would not be told the history of the field, or who it belonged to, or how large the farm was, or how the field looked *before* it was improved, or what it cost to improve it, or whether or not its improvement was profitable to the farmer (*GGL*: 91).

The concerns were the limited concerns of academic experts. Such a gathering, he concluded, revealed problems in the applicability and utility of knowledge "greater than any they had solved":

> How are we to keep knowledge from being bottled up in the bureaus and universities? How are we to get the solutions within reach of the problems? How are we to test knowledge in practice, in the lives of people, as well as in the controlled conditions of laboratories and experimental farms? How do we bring disciplines together? How do we learn to subject technology to the measurement and restraint of cultural value? (*GGL*: 91-2)

Making agricultural science truly practical meant not just insuring that solutions were affordable by those who were to use them, but that they reinforce the culture in which they were used: "a good solution must be in harmony with good character, cultural value, and moral law" (*GGL*: 145). One had to consider how a solution would affect the practice of the larger disciplines of community and faith.

To the extent that university knowledge can supply such solutions, that knowledge must be interdisciplinary, at minimum; there must be research programs in which "economic . . . political, social, and cultural effects" are considered. But there still remains a tension between the idea of farming on the basis of general knowledge and Berry's twin concepts of education as growing into one's place in a culture and of the uniqueness of each farm-farmer relationship. A wise use of scientific knowledge is still possible; such knowledge should promote "the greatest possible technological and genetic diversity, in conformation to local need, as opposed to the present dangerous uniformity in both categories" (*UA:* 221). It "would enable the diversification of economies, methods, and species to conform to the diverse kinds of land. It would always use plants and animals together. It would be as attentive to decay as to growth, to maintenance as to production" (*UA:* 89). In this scheme the agricultural scientist would be the crafter of possibilities developed in the context of the lives of real farmers, and not the purveyor of single truths, "ready-made thoughts," unsusceptible to criticism or modification and hence tending to induce standardized ways of farming. A greater sensitivity to the practical problems of farming could be maintained if there were farmers on agricultural school faculties, just as medical schools and law schools have practicing doctors and lawyers on theirs. In such a case

> the professor would be in a position to "take his own advice before offering it to other people." . . . Professors might again become people of experience rather than experts. They might again be able to apply their learning to the small problems of ordinary people (*UA:* 222).

Berry's outlook is not anti-scientific or anti-technological—he is too much the practical farmer and concerned citizen to revel in ineffable subjectivity. He does not argue categorically against new agricultural chemicals or machines, but worries that we are relying on these as substitutes for restraint and commitment. Such "solutions" may well have unfortunate consequences we cannot yet recognize, but more importantly they simply do not address the cultural roots of the problem.

Berry—Holders of Knowledge

It should be clear by now that in Berry's perspective the most important agricultural knowledge is the farmer's knowledge. The knowledge that comes from colleges and books, no matter how inter-disciplinary, will never match the range of concerns and constraints on the decision-making of the individual farmer. Because it is general knowledge, it will be inapplicable to any particular farm, each of which will present unique characteristics that the wise farmer will have to take into account. Texts and experts

> can be useful to this [the farmer's] mind, but only by means of a transla-tion—difficult but possible, which only this mind can make—from the abstract to the particular. . . . To the textbook writer or researcher, the farm—the place where knowledge is applied—is necessarily provisional or theoretical; what he proposes must be found to be *generally* true. For the good farmer, on the other hand, the place where knowledge is applied is

minutely particular, not *a* farm but *this* farm, *my* farm, the only place
exactly like itself in all the world. To use it without intimate, minutely
particular knowledge of it, as if it were *a* farm or *any* farm, is, as good
farmers tend to know instinctively, to violate it, to do it damage, finally to
destroy it (Berry 1984: 28).

If the farm is not too big, the farmer will be able to know it as
intimately as parents know children; knowledge will come from and
sustain a commitment to work the farm as patiently and gently as parents
would work with their children. But there is still another reason for
locating the most important knowledge in the farmer: the overwhelming
complexity of the ecological interactions on a healthy farm surpasses the
capabilities of any formal knowledge-gathering and decision-making
system. Only a concerned person, employing a combination of close
observation, reason, tradition, instinct, and intuition can manage this
complexity sensitively and effectively. Such an approach necessitates
small farms:

A mind overloaded with work, which in agriculture usually means too
much acreage, covers the place like a stretched membrane—too short in
some places, broken by strain in others, too thin everywhere. The
overloaded mind tries to solve its problems by oversimplifying itself and
its place—that is, by industrialization. It ceases to work at the necessary
likenesses between the process of farming and the processes of nature and
begins to order the farm on the assumption that it should and can be like
a factory. It gives up diversity for monoculture. It gives up the complex
strategies of independence (the use of manure, of crop rotations, of solar
and animal power, etc.) for a simple dependence on industrial suppliers
(and on credit) (Berry 1984: 24).

Berry—New Knowledge

For many of those involved in agricultural controversy the sources and
direction of innovation are of central importance, but for Wendell Berry
they are not. In his view we vastly overestimate the importance of new
knowledge due to our adherence to the prevailing illusion that

the future will be *entirely* different from the past and the present, . . .
because our vision of history and experience has not taught us to imagine
persistence or recurrence or renewal. We disregard the necessary
persistence of ancient needs and obligations, patterns and cycles, and
assume that the human condition is entirely determined by human *devices*
(*D&H*: 148).

Frequently, we look to novelty because we don't care for the discipline old knowledge requires. But new knowledge uncontrolled by old disciplines is apt to be dangerous. It results in "a profound disorder in which men release into their community and dwelling place powerful forces the consequences of which are unknown" (*D&H*: 127).

If production of new knowledge is to be regarded with the greatest circumspection, transmission of old knowledge is of central importance (*D&H*: 103). The culture we must acquire to live responsibly is embodied in our history; we must assimilate and in the process reproduce that culture. We cannot do so through watching, reading, taking classes, or contemplating; we must act in the world. Berry makes much of the concept of apprenticeship with its connotations of learning not just techniques, but standards, disciplines, and satisfactions of a craft. But because it involves a judgment of when and how to act, education cannot be reduced simply to knowing how to do things. Even when one has learned how, there is no simple rule to determine when and where; one learns to live responsibly by practicing responsible living.

Berry—Knowledge and Goodness

Unlike both Battelle and Walters, Berry does not set questions of knowledge apart from questions of morality. For him the best knowledge is situated knowledge and hence entails moral responsibility (Figure 9.1). To confine oneself to so-called "objective" knowledge is a retreat from responsibility that must ultimately fail, for in using that knowledge we must apply it to something, and in application we confront real conditions and consequences that cannot be entirely foreseen. The inadvertence of these consequences does not lessen our responsibility for them. Berry writes:

> Under the discipline of unity, knowledge and morality come together. No longer can we have that paltry "objective" knowledge so prized by the academic specialists. To know anything at all becomes a moral predicament. Aware that there is no such thing as a specialized—or even an entirely limitable and controllable—effect, one becomes responsible for judgments as well as facts (*UA*: 47-8).

Figure 9.1 What Knowledge Matters Most

Berry	knowledge of how, when, whether
Agriculture 2000	knowledge of what can be done
Walters	knowledge of hidden secrets

As an example of the inescapability of responsibility, Berry considers the development of implements of cultivation by early humans. Developing digging sticks, or later stone or metal tools, was an advance in technical knowledge. The accompanying "moral predicament" was that one was suddenly faced with questions of when and how to use tools (or, in Berry's idiom, under what disciplines to use them) to insure that the land's productive capacity was not diminished. Technical skills thus brought with them the need for skills of responsibility, which were embodied in ritual and culture. In this early period, argues Berry, the need for restraint was felt directly and immediately—each person's livelihood was tightly bound to home and community and "certain moral imperatives . . . [were] therefore pragmatically essential." If one did not act responsibly toward nature and society one starved. The domestication of draft animals strengthened these disciplines. It required skill, wisdom, patience, and humility to keep animals healthy and induce them to work. By contrast, the farmer who uses engines powered by non-living energy sources has lost the immediacy of restraint: "a machine is directly responsive to human will; it neither starts nor stops because it wants to. A machine has no life, and for this reason it cannot of itself impose any restraint or any moral limit on behavior" (*UA*: 92-3).

Because we no longer see so clearly the need for responsible use of machines, we are apt to imagine that their design and use is relatively free of moral complications, that the knowledge they embody is good in and of itself. The alternative to this amoral conception of technical knowledge is the development of human skill, which Berry defines as

> the enactment or the acknowledgment or the signature of responsibility to other lives; it is the practical understanding of value. Its opposite is not merely unskillfulness, but ignorance of sources, dependences, relationships.
>
> Skill is the connection between life and tools, or life and machines (*UA*: 91).

Reorganization of the institutions of knowledge, recognition of the demands of responsible human living, and a re-articulation of the good our actions are intended to enhance will allow us to construct the kind of whole knowledge we have so little of now.

Walters

Walters—What Is Knowledge?

In contrast with Wendell Berry, Charles Walters jr does recognize a special place for knowledge. Progressive social change may come through action, but the ability to act effectively depends on the integrity of a body of knowledge. As we have seen, Walters is concerned with our failure to recognize important truths, such as the eco-agricultural laws of plant and human nutrition. This knowledge is empirical knowledge, but once we have acquired it we can make it a guide for action by deducing from its general laws the appropriate response to particular situations. Unlike Berry, Walters assumes that the universe is a rational and intelligible construction, though one far more complicated than we currently appreciate. Like Berry, he fears that we may be losing important insights laboriously acquired over the centuries by working farmers.

For Walters agriculture will be successful and sustainable only when informed by a thoroughgoing ecological understanding—when the manifold bits of right answers have become part of a system of general knowledge. He calls this "high science" and distinguishes it from the pseudo-investigations of current agricultural researchers, who make misleading and irrelevant observations because they lack an overarching ecological perspective and therefore fail to ask the right questions (*Acres* 1/81: 3):

> Most of what is generally called the "scientific system" is not science at all, but merely procedural. The procedural aspect calls for setting up experiments that eliminate other possibilities, or it deals with making instruments that enable the investigator to find what he is looking for. The backbone of the scientific system has to do with asking the right questions. A scientist can only ask the right questions after his life has absorbed the experiences that lead him to a vision of the Creator's handiwork, hence the right question. . . . When science falters, it is because no one is asking the right questions (*AP*: 416-17).

With the right questions will come just and humane social science as well as correct natural science, for in Walters' view truth and justice are fortuitously interrelated: the fairest and best way for humanity to be is already set up in the natural order and would be apparent if we understood nature properly. Fuller knowledge of nutrition and of the rest of the "everything-relates-to-everything-else equation called eco-agriculture" will also transform human nature. In Walters' view we are far from reaching the limits of the human definition because we are guilty both of perpetuating the malnourishment and poisoning of our own bodies, and of failing to realize what we might become. Readers of *Acres* can learn how to use the wisdom of the builders of the pyramids or the ancient round towers to achieve levitation and enhance fertility. They will also discover that we are less than we might be, owing to the damage done to us by poor food, fluoride in our water, pesticides in our environment, white bread, pasteurized milk, electric blankets, and the steel shells of automobiles (*Acres* 1/82: 10, 2/82: 24-6, 3/82: 4, 4/82: 10, 6/82: 39, 7/82: 12-13, 8/82: 8, 28, 9/82: 32).

All these examples suggest that the Creation is a treasure to which we as humans have access through knowledge. We are unlikely to discover the keys to this treasure through conventional science for it is too much linked to maintaining the existing errors; nor will the keys be discovered by Berry's farmers, for they will be so busy behaving responsibly within what they mistakenly think are the limits of the human definition that they will have little vision of what more human life might be.

Walters—Institutions of Knowledge

Walters' vision of what agriculture might be is more dependent on knowledge (especially scientific knowledge) than Berry's and his assessment of what is wrong with current institutions of knowledge is correspondingly more prominent.

Conventional knowledge production is distorted in three ways, Walters believes: first, by the ties of researchers to vested interests that exploit farmers and perpetuate petroleum-based agriculture; second by the organization of research, which leads to misleading fragmentation of truth; and third by the confusion of normative with descriptive knowledge, which is a consequence both of the dominant philosophy of applied science and of the application of research to policymaking. We consider the first two here, the third below.

In the preface to the *Acres U.S.A. Primer*, Walters and Fenzau write: "There is nothing occult about science. And we add, there is nothing occult about the rationales constructed by the grant receivers to protect the commerce of their patrons" (*AP*: xii). In a nutshell, Walters explains in the same way the failure of conventional agricultural scientists to recognize either the economic rationality of the NFO or the scientific rationality of eco-agriculture: they have been bought off. There has been a great deal of "scientific gangsterism," of discovering only the truths that money can buy: "As they used to say at Iowa State, 'For $100,000 we can prove anything'" (*AP*: 167-8).

But as with the 4-H movement and the Farm Bureau (see Chapter 7), the corruption has often been insidious. Researchers have maintained good will toward farmers, consumers, and nature all the while their conceptions of reality, social possibility, and scientific method were being narrowed and redirected by funding sources.

The key changes took place after the second world war. Prior to 1950, he argues, agricultural scientists were progressively unraveling the enormously complex biochemical processes of the soil. Beginning with the work of Serge Winogradsky, bacteriologists had begun to understand the complicated requirements of nitrogen-fixing bacteria. Soil scientists, W.A. Albrecht of Missouri in particular, had developed an understanding of fertility that went far beyond Liebig's NPK (nitrogen, phosphorous, and potassium) concept. They took an interest in the trace elements and the ways in which weak humic acids made nutrients available to plants. Periodically development of cheap fertilizers threatened this research program, but it was only with the opening of middle eastern oil fields in the early 1950s that "oil company technology conquered all—USDA, the colleges, the extension workers and farmers. . . . advanced science lost its way in favor of a primitive form of soil chemistry" (*CEA: 72*).

As big oil sought to develop a market in energy-intensive agriculture the guidance that scientists provided farmers had to be revised.

> Long accustomed to buying politicians like bags of popcorn, the cartel managers proceeded to buy up education. The grants—that is to say, the acceptable bribes—were so subtle even the recipients didn't understand how money influenced both the findings of science, and criticism thereof. Departments in universities needed money. It was kept coming only when results were suitable and usable. Unsuitable results didn't even rate publication. Obstinate professors were simply drummed from the corps. In order to synthesize honestly, the common denominator of the academic community became *simplify*. Nothing could be simpler than magic dust styled N, P, and K. Nothing could be more easily sold than university blessed fossil fuel technology as long as farmers believed in the professors (*CEA: 73*).

Walters is not impugning the motives of researchers here, but he is suggesting that the lure of a shortcut to short term productivity duped them into giving up their attempt to understand scientifically how fertile soils really work. In his view the range of conditions that affect yields has not been appreciated by conventional agricultural scientists, with the result that they are making mistakes as glaring as if a chemist were to use dirty glassware or a bacteriologist non-sterile media. A true science would show that nature is vastly more complex than scientists (applied scientists in particular) realize. But unlike Berry, Walters does not see this complexity as indicative of the ultimate inapplicability of the general knowledge of science to the singular needs of particular farmers on particular farms. For Walters a better and purer science really is the key to a vastly better world, but we shall not get that science without a thorough cleansing of the universities. We shall have to find neutral sources of funding and we shall have to reject a great deal of what we have been mistakenly taking as truth for the last forty years (*AP*: 95-6).

Walters—Holders of Knowledge

The question of the utility and integrity of the farmer's traditional knowledge is a central one in contemporary agricultural controversy. For Berry the best knowledge is farmers' knowledge. Walters, on the other hand, sees most people, including most farmers, as not astute enough or as insufficiently attuned to nature to practice ecologically sound agriculture on their own. That farmers have enthusiastically embraced oil-based agriculture shows how little they understand: farmers "may still remember that nature created life, but they think the test tube and the fossil fuel factory have vacated nature's rhythm of life and death" (*AP*: xiii). They have forgotten how to read the landscape and "some want to know little more than what the labels say" (*AP*: 109). Farmers' knowledge of nature manifests itself only when they complain about the weather, yet with a little effort they could take intelligent charge of all manner of other natural processes and increase their productivity enormously:

> In the coffee shops farmers like to talk about drought, about too much water, a pastime known as the *blame the weather* syndrome. Fertility, tilth, soil physics, all would be better topics. . . . Farmers probably would turn to such topics if someone led the way" (*AP*: 229).

Thus, far from embodying the accumulated folk wisdom that has sustained rural people since time immemorial most of America's farmers are as dull as the rest of the population.

Walters is also critical of alternative agriculturalists who fail to appreciate the scientific bases of eco-agriculture. These are "organic folks" and their beliefs (about composting, for example) are "folklore," a derogatory term in his lexicon. He is glad they have rejected conventional agriculture, but bothered that they have taken up an equally simplistic dogma (*AP*: 400, 201). They may stumble onto ecological truths, and practices consistent with those truths, but so long as they insist that old ways or natural ways are *the way*, they will fail, and more importantly, fail to demonstrate to conventional farmers (and to agricultural researchers and policymakers) that ecological forms of agriculture are viable and desirable. (*AP*: 247, 221, 201, 203; *Acres* 1/81: 3).

Having rejected conventional agricultural science as corrupt and farmers' knowledge as inadequate (or, in the case of organic farmers, naive and unsystematic), Walters turns to the consulting firm both as the source of the best knowledge and as the best means of getting it into the hands and heads of farmers. Consultants will recognize truth and deliver disinterested advice to farmers simply because they are free from institutional or ideological biases.

Consultancy figures prominently in *Acres*. The coauthor of the *Acres Primer* is C. J. Fenzau of Advanced Ag Associates-West and one of the main markets for the *Primer* is the clientele of Eco Soil and Water, a Nebraska consulting firm with franchises throughout the high plains (*AP*: 128-9; *Acres* 9/82: 3). The services consultants offer range from on-farm analyses to seminars on "first principles" (*Acres* 7/82: 24-5). These consultants embody the new eco-agricultural science that Walters sees replacing the simplistic dogma of the land grant universities. For example, Don Schriefer, a "bio-ag" consultant interviewed in *Acres*, outlines novel perspectives toward the soil (viewing it as a factory) and toward decomposition (a concern with "quality annual decay"), and suggests new ways of posing central questions of agriculture: "If we could talk to the plant, what are the basic things it would tell farmers to do?" (*Acres* 7/82: 24-5).

At times the approach appears condescending toward farmers. The attempts of a seaweed fertilizer firm, Micro Ag of Washington, to teach farmers the benefits of spraying the algae chlamydomonas on wheat fields are described thus: "Stuffed toy models and posters of 'Sammy, the Chlamy' also lead Micro Ag spokesmen into telling farmers about the electrons from the atom of another element in order to get eight electrons for its outer orbit" (*Acres* 7/82: 27). From the *Acres* perspective such educational efforts, however undignified, are necessary because farmers ignorant of fundamental biochemical processes are spending enormous amounts of money on destructive and dangerous chemicals. For Walters it is not only legitimate to import expert knowledge to the farm, but necessary. To depend on the native ecological wisdom of farmers is to stumble along destroying nature and ourselves, inadvertently no doubt, but no less irreparably.

Walters—New Knowledge

In contrast to Berry, Walters is centrally concerned with new knowledge. The focus of eco-agriculture is on finding the true science of plants, animals, and soils. Hence questions of where current science has gone wrong, how to correct its course, and what sort of process enables us to discover truth are central. Earlier we quoted Walters' view that the key to the "scientific system" was asking the right questions. But what are the right questions?

In Walters' view conventional agricultural scientists have committed a major methodological error in practicing "single factor analysis," that is, in attempting to measure the effects of one variable while holding others constant (*AP*: 98). The results of such a practice are bound to be useless, he argues, since it is fundamentally inconsistent with the "everything-re-lates-to-everything-else equation called eco-agriculture" [1] (*AP*: 94). He admits that the alternative, a thoroughgoing holism in research, is difficult. Yet we fool ourselves if we think single factor analysis and the specialization that accompanies it is an adequate substitute:

> man's puny mind can't comprehend the many ramifications involved here [in liming and the cycling of organic matter] without fencing off each topic and handling it as if in isolation. We will do this, of course, but always at the cost of some misunderstanding. . . . Single factor analysis . . . is, in any case, the curse of the amateur. We realize that the single variable has been and remains the lodestone of much college research, but it has all too little to do with production on a living, vibrating, dynamic farm. It ruins true science by making specialists out of young minds, killing off their imaginations (*AP*: 98).

The objection may seem similar to Berry's: the farmer is of necessity a holist—"sooner or later the farmer has to bring it all together so that his mind can handle the whole of it all" (*AP*: 111). Yet in Walters' holism humans participate only as components (if very important components) of physical, chemical, and biological systems. It is not a holism that responds to, or is driven by, human character and culture.

All in all, Walters portrays agricultural scientists as methodologically naive, as using simple methods and making unwarranted simplifying assumptions to study that which is enormously complex and variable. He admits that even his sophisticated system of soil audits does not reflect the true state of the soil, and that ultimately one comes up against an uncertainty principle that makes definitive soil analysis impossible:

> How accurately do soil audits reflect the extremely variable conditions that changing temperatures, growth patterns and physical conditions—tilth, weather, toxicity—impose? After all, the results of a soil audit detail a condition that no longer exists. Moreover, grinding soil samples imposes a condition that never existed in the first place (*CEA*: 64).

We are left then with an agricultural science both incompetent and misdirected. Legislation to require research on organic agriculture will not help if scientists aren't competent: "We do not think most chemically trained 'scientists' can be flexible enough intellectually to keep an open mind" (*Acres* 8/82: 5).

A new science will require a fundamental change of world view. We must see nature differently and ask different questions of it. The answers we seek may in many cases be readily accessible yet without a new perspective we will not recognize them. We will not find this new perspective in existing institutions; both the exploited and the exploiters in our society have been too long dulled by ideology (and poor food) to recognize what they are looking at.

An example of new perspectives leading to new questions is the observation that since not all members of a given species are attacked by a pest or disease infestation there must be unrecognized determinants of resistance possessed by only some members of the host species. Walters suggests that trace nutrients in the soil are essential for this resistance (*AP*: 270-1). Here an interesting hypothesis about plant nutrition arises by granting greater status to a phenomenon hitherto taken for granted. We open ourselves to such insights by thinking in terms of new metaphors. Walters recalls Albrecht's recognition that the soil was to the plant as the stomach was to the animal, an insight that led to development of an agriculture that did not destroy digestive processes of the soil. For Albrecht, this observation was so obvious that "even children" could make it (*AP*: 119). Yet it was an act of genius because it represented an escape from conventions that still bound the minds of most scientists.

Walters—Knowledge and Goodness

For Walters, objective knowledge of nature and of society liberates us from the constraints of poor health, unjust institutions, and unsound agriculture. These truths will liberate us because the plan of creation is a rational one that includes social and ecological patterns that are good for us and to us. At present we can recognize this good but know too little of nature or social relations to achieve it. Hence the moral significance of knowledge is that more knowledge (real knowledge not ideological deception) brings more morality—morality, like knowledge, being something we can accumulate and its production something we can maximize. To know is not enough, however; what *is* is not necessarily what's right. We must find the meaning of what we discover by learning to read the normative signals nature and society continually broadcast. When we look around we see a mass of problems and a sprinkling of hints of how to correct them, but we must distinguish the effects of our own inept management of things from the workings of natural laws.

These issues are especially important in Walters' discussions of social science. Much of the anger in *Angry Testament* and *Holding Action* is with agricultural economists and sociologists who have drawn prescriptions for policy from their descriptions of farmer and consumer behaviors. Their descriptions have been taken as indications of what must be and have been allowed to take on normative significance and dictate policy. The process creates self-fulfilling prophecies, since if we treat society according to a particular notion of how we think it will behave, we are setting up structures that facilitate that behavior. In Walters' view this approach to applying knowledge is completely wrongheaded; rather than use such descriptions to justify what is, we should use them diagnostically to discover what is wrong and how to fix it.

Walters' exemplars of illegitimate normative inferences are a number of reports on American agriculture published in the early 1960s: "An Adaptive Program for Agriculture" by the Committee for Economic Development (CED), Michigan State University's "Project '80," and the report on "Problems of Progress in Agricultural Economics," written by Dale Hathaway as a member of President Kennedy's Council of Economic Advisors. These analyses were concerned with surplus farm production. They favored reduced farm prices along with the disappearance of many smaller farms. These measures would reduce surpluses, help the country compete in food export markets, and stimulate industry by both freeing up labor and leaving consumers with extra money to spend.

None of these reports "spoke pure economics," Walters writes. Rather, they masked political interests as economics (*HA:* 86; *CEA:* 40-2). They were not speaking "pure economics" because they had failed to take into account the causal economic law that real wealth originates in agriculture and that therefore low farm prices destroy prosperity. Unjustified interpretations had led not only to unwise policy, but also to a moral dilemma of justifying the detrimental effects the policy had on farmers. The abstract rhetoric of social science was used to gloss over the individual tragedies the policies caused. Walters wrote of the CED report:

> This was all polite stuff, neatly turned academic language for the greatest class exploitation in history. . . . And the economists could see this as merely "creating serious problems for some rural communities," when in fact overurbanization was not a wry epigram, but an unholy fact, a stinking cancer, a monument fouled with the odor of human defeat (*AT:* 33).

Here the language of neutral description hid the moral outrages that were being perpetrated. Walters labels this kind of academic rhetoric "frustration economics" or "frustration sociology" on the grounds that it uses knowledge to frustrate the achievement of justice (*AT:* 203; *CEA:* 41; *HA:* 140).

Battelle

Battelle—What Is Knowledge?

The perspective of *Agriculture 2000* overlaps with the perspectives of Berry and Walters, yet differs significantly from each. Like Berry, *Agriculture 2000* is concerned more with knowledge of how than knowledge of what—how to improve plant and animal stocks, increase soil fertility, control pests, feed livestock, cultivate land and harvest crops, satisfy demands of consumers and rural residents, and so on. Again, like Berry, it sees this knowledge as situation-specific. The knowledge of agriculture is of how to do things efficiently in a given economic environment. In our discussion of nature we pointed out that it is hard to find in the book an image of a transcendent nature that limits what is possible and causes what happens, such as we find in Walters' works.

In *Agriculture 2000* both limits and causes are economic. As what is "natural" becomes increasingly identified as what is "economic," the knowledge of nature (including human behavior and social institutions) becomes increasingly a knowledge of how production, distribution, and consumption problems are solved under different sets of economic conditions. For example, knowledge about the salt tolerance of food crops is constituted in terms of economic conditions and constraints: salt-tolerant cultivars may not now exist (we have no knowledge of their existence) but given sufficient incentive we will invent them and will then have knowledge of them. For Walters (and to some extent for Berry) there are natural limits which we can know and which, in many cases, we are close to reaching. For them the problem of salinization is whether nature includes salt-tolerant food crops in its storehouse of possibility. Both doubt that nature has unlimited capacities of this sort.

The authors of *Agriculture 2000*, on the other hand, refuse to let the range of possibility be held hostage to claims about nature. Perhaps there are natural laws that limit possibilities, but we only come to an approximate knowledge of them by trying to solve economic problems. And our knowledge will always be tentative for with greater incentives we may do what now seems impossible. Fifty years ago the limits of reduction in physics lay with the nucleons of the atom, fifty years before that the limit was the atom itself. Money for research overcame both obstacles, hence it is unwise to assume that nature imposes absolute limits on what can be known or done. What this suggests is that knowledge of *how* is the only ultimate knowledge we have. For example, laws of plant physiology that lead us to believe that salt-tolerant cultivars are impossible must be recognized as subordinate to technical knowledge: if we manage to invent salt-tolerant plants, we will have to revise what we thought were laws of plant physiology. This view also suggests a definition of knowledge: "all we know is what works." Ultimately knowledge is a matter of what we can do; it literally is power. We can organize it and make up theories and deduce likely possibilities, but we can't know if our deductions are true until we put them to work. And if they do work their status as fact will not depend on the theories that suggested them.

Battelle—Institutions of Knowledge

Like Wendell Berry, the authors of *Agriculture 2000* are critical of knowledge producing institutions: their knowledge is too general or too specialized for specific situations. But unlike Berry, for whom the possessors of the most important agricultural knowledge are and always will be good farmers, *Agriculture 2000* sees knowledge as embodied in "smart" instruments and machines and in the managers who program and use them. The knowledge itself will exist in the form of solutions to complex optimization equations with large numbers of variables. Knowledge generating institutions such as universities can contribute to this kind of agriculture, though they have not always done so very well in the past.

As agriculture becomes more exclusively a business and less a way of life, farming knowledge will increasingly become the possession of a class of professional managers. Those who go into "agribusiness" (which as used in *Agriculture 2000* includes farming) will have to have a "business knowledge, a business orientation, and devotion to proven business methods, . . . necessities that too many specialists and technicians do not have today" (Battelle 1983a: 80). Even scientists will be required to be adaptable and practical. For example, it will be "extremely important that plant scientists . . . have good business and management training. . . . These specialists will not be confined to the research lab, but will be expected to counsel with customers and help resolve product complaints" (80). There will even be a specialty for the person who adapts knowledge to specific situations, who will be known as the "management information specialist":

> Such people must understand what information management needs (not necessarily asks for) and how those data can be obtained from data

processing and in what form, and they must understand financing and why financial considerations could preempt all other considerations. This requires a well-educated, diverse person, capable of understanding several disciplines at various management levels (80).

As in Berry's writings a practical attitude is in the forefront here. How many degrees one has doesn't matter, since the real world is quite different from the world of textbooks. Yet unlike Berry, the Battelle authors believe that despite this variability we can and should take a reductive approach to farm management.

> The importance of managerial skills to a farmer's welfare has increased as farms have increased in size, as purchasing and marketing . . . increased in complexity, and as the level of technology used on farms has advanced. Although there is general agreement that the management process plays an increasingly important role in the success of a farm operation, farm management teaching and guidance has remained in its infancy. The human factors which contribute to the managerial process and behavior are often difficult to measure and to transfer. Its [sic] often easier to evaluate the outcome of managerial skills than to measure the skills themselves (7).

From Berry's point of view the reason "human factors" are so hard "to measure and to transfer" is precisely that they are human, and the reason that it is so difficult to reduce good management to a science is that it only exists in individuals who are products of a healthy farm culture. There is no point in trying to manufacture such a science. For *Agriculture 2000*, there is. Farming has become so competitive that efficient management is necessary for survival. Sound management involves eliminating sources of unpredictability, and humans are one of those sources. Hence, while Berry sees the integration of knowledge as leading to an acceptance of the variety of human concerns and habits, the Battelle report suggests that with sufficient cultivation of the science of farm management, we will be able to understand, take into account, and thereby circumvent or neutralize the inefficiency inherent in human factors. Farming will present decreasing risk to the capital invested in the farm, the more the range of unpredictability can be confined. The report agrees that we need flexible and adaptable knowledge, but we need it to insure that goods and services are produced and delivered as efficiently as possible, in spite of the inefficiencies we humans may sometimes be inadvertently causing.

Battelle—Holders of Knowledge

One key difference between the Berry and the Battelle concepts of flexible and situation-specific knowledge is that the Battelle report insists that knowledge be public knowledge, readily available to trained farm managers, rather than the tacit, indigenous knowledge of farmers (see Figure 9.2). In the future one of the main ways this knowledge will be made available is through development of expert farming systems that will reproduce the sensitive judgments good farmers have always made. *Agriculture 2000* foresees computerized systems which, for example, will mix feed rations to suit the peculiar biological situation of each dairy cow (142). In another new system, "custom prescribed tillage,"

> requirements for the seed bed, the root zone, and pest control for soil and water conservation will be specified. The next step is to use the available information on soil dynamics, machinery, economics, and climate to prescribe components of the tillage system and how they should be used. The specific machines to be used will be defined, as well as the sequence and timing of their use. With this system only the appropriate amount and type of tillage will be used (124-5).

This new approach is distinguished from the more familiar "reduced tillage" in that "custom prescribed tillage . . . is a dynamic system using information feedback to modify tillage practices through the year. As weather and crop conditions change, the prescribed tillage will also change" (124-5). A third example is of "optimization settings" for tractors:

> The tractor operator . . . will be able to select any one of several optimization modes in which to operate. For example, if there is a storm on the horizon and planting is under way, the operator will select the maximum work rate setting. This system will provide the maximum available production per tractor horsepower at the expense of fuel efficiency. The tractor reliability and life will be improved because the computer will not allow the operator to select settings that could damage the tractor (119).

Figure 9.2   Who Knows?

Agriculture 2000	experts and expert systems
Walters	"real" (independent) scientists
Berry	farmers

In each example, the attempt is to find a far more situation-specific basis for making knowledgeable decisions by basing them not on the contents of general and out-of-date bulletins or textbooks, but on inputs from sensors processed by computers. From Berry's standpoint, these proposals may seem ironic, even ludicrous. "Custom prescribed tillage," for example, may seem like an ersatz substitute for good farming; it is, after all, the farmer who "customizes" techniques to a unique place and time. In the case of the dairy cows, one might recall a time when farmers named their animals and feed was meted out with regard to the condition and even "personality" of each animal. As for optimal tractor settings, one might argue that farmers have always sensed how their machines were performing, and that they do run their machines optimally with respect to their own priorities, which may be complex, varying, and impossible to measure.

Hence it might be asked, "If good farmers possess this knowledge why is it necessary to inscribe it in sensors and computers?" To answer the question we must draw on several of the report's assumptions about future agriculture.

First, its assumption about farm structure. *Agriculture 2000* foresees larger farms run by fewer humans, a concentration dictated by economic forces. Even the wisest farmers will find it impossible to keep track of the details; one can think of only a limited number of cows as individuals and can give particular attention only to a limited number of acres. (For Berry these were precisely the reasons we must restrict farm size.) Yet once we have developed sensors attuned to what good farmers were attuned to and programmed computers to process their inputs, there would seem no limit to farm size, and we will be able to have large farms sensitively managed. And we will need to do that, *Agriculture 2000* points out, since as the number of farms declines there will be fewer farm-raised and farm-trained workers to staff agriculture-based industries (79).

The structure of future agriculture is only part of the answer, however. It explains why we will rely more and more on electronic information gathering, but not why we will also use automated decision-making systems. Yet that is what is being suggested: custom prescribed tillage "prescribes" and insures that the "appropriate amount and type of tillage will be used." There is little need for the farmer as intermediary between information and action, and even a possibility of conflict between the farmer's judgment and the computer's prescription. Consider, for example, the case of the automated tractor. In the event of an oncoming storm *Agriculture 2000* envisions the farmer setting the tractor on a rapid-work mode in which it would run as rapidly as possible without damaging itself. Yet in such a situation a farmer might judge it wise to risk harm to the tractor, say in order to get out of the way of a tornado.

A final rationale for the rejection both of tacit knowledge and of subjective decision making is financial: modern farm machinery is so costly and so often owned by others than the operator that it may seem desirable to protect it from human error. The "smartness" in machines is thus a kind of investment insurance.

The inscription of knowledge and relocation of decision making also reflects assumptions about the sources of knowledge. Although the knowledge of smart systems resembles what Berry's good farmers are said to know tacitly, it is not a formalization of their experience but the end product of science. What flexibility these systems provide will come therefore from the comprehensiveness of the science on which they are based and from the number, distribution, and sensitivity of sensors that provide the inputs.

Battelle—New Knowledge

At several points we have noted in *Agriculture 2000* a sense in which technical problems are already solved. Since knowledge is only know-how, and since the appropriate technique is determined by economic conditions, all knowledge already exists in a potential sense—it waits only for economic conditions that will make it worthwhile to develop and apply. This perspective is clear in the report's discussion of new technologies such as robot farming:

> Robot farming represents one of the most futuristic ideas for agricultural mechanization. The proposal is highly conceptual, and would require many years and millions of dollars to implement. Successful development will hinge upon the availability of research funding (133).

Together, "futuristic" and "highly conceptual" suggest that the technology is still at a "drawing-board" stage—theoretically a feasible technology that is yet to be proven. The end of the passage makes clear, however, that whether or not robot farming machines will be developed is not a question of what is possible in nature, but of how much money, and hence effort, is allocated to the project.

The report is almost always confident that apparent barriers to knowledge will be overcome in the course of research, frequently by means of a "breakthrough." They write, for example, of hog production: "another breakthrough along with artificial insemination is needed to improve conception rates" (160). Or of development of nitrogen fixing corn varieties: "commercial corn production exhibiting nitrogen fixing characteristics must await several technological breakthroughs before success can be achieved" (53). We have already noted their confidence that new salt-tolerant crops "will allow" farming of saline soils (52).

The term "breakthrough" is key here in sustaining the guarantee of progress. In the quoted passages a "breakthrough" is a major scientific advance that comes about sooner or later according to the effort expended. "Breakthrough" is a common metaphor in science, yet it is used in at least two distinct senses. One is "serendipitous discovery," an accidental juxtaposition of ideas and experience that generates a radically different picture of the world and direction for research: the musing Newton struck by a falling apple, Darwin's reading of Malthus at the crucial time, Pasteur studying the loss of virulence of a batch of chicken cholera virus. Here great scientists make unexpected inferences from close attention to detail; big budgets don't matter. The discoveries may in fact be quite simple yet it is only rarely that the light goes on and we find ourselves to have broken through the veil of ignorance. Breakthroughs of this type are impossible to predict. Presumably they will continue but under what circumstances we cannot say. This image is close to Walters' view that we not plod on with old ways but break through to a new way of regarding and understanding nature.

Since there is in *Agriculture 2000* no concept of nature as an entity out there waiting to be discovered, the inevitable "breakthroughs" in the text belong to a quite different metaphorical web. The "breakthrough" resembles a breakthrough in battle, like the vision that made possible the Battle of the Somme in World War I. Allied generals determined to end the stalemate of trench warfare by making a "big push," an offensive so massive nothing could withstand it. Immovable obstacles could be moved if only enough force were applied. In science this view of breakthroughs belongs to a Baconian tradition: the resistance of nature is to be broken not by insight or genius, but by the effort of ordinary, if well organized, workers.

There are important differences in these notions of breakthrough. In the former scientists do not so much *achieve* as *receive* breakthroughs. Enlightenment comes, if it comes, to one with sensitivity, insight, and open-mindedness. In the latter scientists produce the breakthrough by force of action. In the first nature is a mystery whose solution may be revealed to the patient and worthy, in the second a set of barriers to be broken down. The first sense associates science with reflection and inspiration, the second with mobilization of resources and with disciplined collective activity.

Which of these senses of breakthrough is judged appropriate will depend partly on other images—of nature, of how societies work together, of what problems are most critical, and of whose responsibility it is to solve them. For example, if we see nature as ready to provide us with its bounty so long as we are properly humble, we are likely to be impatient with the "hubris" of agricultural science. If, on the other hand, we see nature as undeveloped potential, we are likely to be frustrated by those who urge us simply to contemplate and hope for eventual enlightenment. Which of these senses we use may also vary with the problem at hand. Enlightenment may be the route to the inverse square law, yet be unsuitable for the development of a nitrogen-fixing corn.

Battelle—Knowledge and Goodness

The Battelle report sees knowledge of nature and society as solving moral problems, and moral problems are presented simply as inappropriate ways of posing economic problems. This may seem similar to Walters' framework, but it is in fact quite different. For Walters true knowledge makes a more just order possible, but moral questions (such as whether an action is just) are distinguishable from empirical questions (such as whether or not lack of trace elements causes subclinical malnutrition). In *Agriculture 2000* on the other hand economics is the science of morality as well as possibility. We figure out what is right by translating our options into economic terms and trust to the goodness inherent in economic laws (which are based after all on the pursuit of good) to insure that the conclusion we come to is right.

At first sight it may seem strange to raise questions of knowledge and morality at all here. After all, we have repeatedly stressed that knowledge in *Agriculture 2000* is knowledge of *how* to accomplish a necessary task, such as feeding ourselves, and that the necessity is imposed by the need to survive in a world of limited resources rather than by moral teachings. In other words, we get knowledge to solve problems because we recognize that we must do so to survive. Because the conditions define the problems, and the problems determine the knowledge needed to solve them, questions of whether that knowledge has moral implications need never arise. To raise such issues in any serious way would be pointless, since however we chose to answer them would not have one iota of effect on the problems or the knowledge needed to solve them.

Consider for example the issue of farm land in the southwest. In Berry's view the farming practices of one indigenous population, the Papago, reflect a moral choice, based on a knowledge that was both a knowledge of how, and of when and where to farm. Their agriculture embodies "an entirely competent understanding that Papago land is agriculturally marginal" (*GGL*: 67). In his view, conventional "agribusiness" use of such land (to irrigate it heavily) is based on "an elaborate pretense that desert land is not marginal." But in the view of *Agriculture*

2000 no pretense is required. Irrigated farming meets demand in the most efficient way; it is therefore legitimate, even necessary. The economic calculus then, incorporates consideration of when and where action is appropriate.

The clearest example in *Agriculture 2000* of how economics can resolve moral dilemmas is its discussion of confinement animal-raising techniques. Some argue that such practices violate animal rights:

> In the opinion of the animal rights advocates, many aspects of intensive agricultural systems produce stress on the animal, and are considered to be inhumane practices. (This subject is controversial because livestock producers and scientists traditionally consider the welfare of food animals in terms of their growth rate, efficiency of feed use, reproduction efficiency, mortality, and morbidity.) Stress, however, cannot be measured numerically, and the physiological behavioral measurements used as stress indicators are difficult to interpret. Some scientists feel that livestock in modern confinement systems are under significant stress and therefore animal welfare is impaired. Other scientists believe that animal welfare is not affected unless the traditional criteria of animal performance are significantly impaired (59).

This passage is especially important in that it recognizes and attempts to synthesize two very different perspectives. In a sense it is a microcosm of the way parties to controversy will try to resolve in their own terms problems brought up by others. Here the Battelle authors seek a definition of animal rights that integrates animal rights considerations with other variables in farm management.

The passage interprets animal rights claims in terms of stress and sees stress as a possible factor in animal welfare. The question of whether raising animals in confinement is inhumane is, in effect, translated into the question of whether traditional physiological and behavioral criteria of animal welfare (growth rate, efficiency of feed use, reproduction efficiency, mortality, morbidity) can or should be expanded to include indicators of stress. The authors recognize that this will be difficult; scientists do not agree on the definition of stress, and hence on how to measure it.

Critics of conventional agriculture may ask why the authors take so complicated an approach. Some of them may see "inhumanity" as a matter of one's intentions toward animals, therefore having nothing to do with elusive questions about stress. Others might find it more reasonable to identify animal rights as moral constraints on animal raising that require ethical rather than scientific analysis. The Battelle authors, however, are working within an ethical tradition of utilitarianism that approaches the question of how animals ought to be raised by asking what are the consequences of the alternative ways of raising them and by seeking to maximize good consequences and minimize bad ones. They seek to integrate ethical analysis with economic analysis by relying on a utilitarian approach in both. In their view the most "economic" methods of animal raising will, by definition, bring about the greatest good of the greatest number.

Both from an ethical utilitarian and an economic point of view, it is the need to analyze consequences accurately that leads the authors to attach so much importance to treating animal rights in terms of stress and to consider quantifying stress and treating it as one of the factors in animal welfare. Unless we know what factors contribute to animal well being, we cannot be sure what the consequences of our treatment of animals are. Hence far from being dismissive of those who are concerned about animal rights, the Battelle authors have gone out of their way to take them seriously by seeking to translate such concerns into an idiom that will allow them to be evaluated and their true weight determined.

Ultimately, the concern that moral problems be dealt with quantitatively through science and economics comes from the same sources as the concern that good farming know-how be formalized if it is to have the greatest utility. In both cases the impulse is to put knowledge to work in systems that are as free as possible from human error and idiosyncrasy. Berry, by contrast, would find it quite appropriate that a farmer make decisions according to his or her own self-perceived circumstances and sense of rightness, regardless of how the economic calculations turned out. In a culture which practiced disciplines of responsibility we could be confident that animals would be treated humanely even though the foundations of humaneness would be located in the sensibilities of individuals rather than in an algorithm of animal welfare. But in the view of *Agriculture 2000* to allow such an approach would be to permit economic chaos. Without quantification there can be no sound management and no accountability. There would be no way of rationally justifying investment policy if a key factor of economic feasibility could not be included in one's calculations. Indeed, far more serious than the threat that animal rights advocates might force farmers to make uneconomic choices is the possibility that their concerns might undermine our ability to predict, to plan, and hence to manage with any degree of confidence. And without reliable management, everyone's livelihood becomes subject to "outrageous fortune."

Notes

1. Not only does single factor analysis not lead to truth, it actively leads to error. Walters, like Rodale, argues that studies comparing organic and conventional agriculture have produced meaningless results by focusing on single factors. For example, despite the fact that most organic farmers and gardeners recognize that organic processes will work well on fields which have long been conventionally farmed only after a conversion period during which soil organic matter builds up and soil structure improves, some university researchers have tested organic practices on plots "worn out" by experiments on conventional agriculture.

> Research people made neat tabulations, replicated their findings in worn out plots of ground, and announced a new scientific finding to the world. Needless to say, if such tests had been conducted on soils managed by organic gardeners or good eco-farmers, they'd have gotten no response at all [i.e., no negative result for organic farming]. . . . Rodale and the organic gardening people have been right all along. Presence or absence of organic matter in the soil determines how these college experiments come out (*AP:* 168).

10

Social Order

Images of social order figure in contemporary agricultural controversy in many ways. They are most visible in relation to questions about the structure of agriculture: about how many farmers there ought to be, how land and other assets should be owned and distributed, and what relations should exist between farmers and other sectors of society—consumers, food processors, the banking industry, and so forth. The images we describe here represent important if unrecognized differences among the parties to agricultural controversy. We focus on five questions of social order.

The first is what is *society* as it ought to exist and as the term ought to be used? Is it a community, a just balance among competing groups, or a bustling economic machine?

Second, how ought we (and others) to identify ourselves and understand our social roles? Should we think of ourselves as multi-faceted persons, or as members of a particular occupational, national, or ethnic group, or simply as individuals participating in an economy and a democracy?

Third, how do we think about others, especially those whom we disagree with or feel in competition with? Are they much like ourselves or quite different? Do we divide groups in society into subjects and objects, "us" versus "them"?

Questions four and five open a theme that we continue in the next chapter on Praxis. It concerns broad issues of why things happen—what goes wrong in society, what governments can and ought to do, and how social change comes about. Question four concerns the origin of social problems. Are they rooted in nature—for example, in the unequal distribution of natural resources—or are they the results of policy errors, or are they simply the totality of problems of individuals?

Fifth, what is the role of state, national, or international political institutions and processes in solving these problems? Ought we to look to or away from government in seeking solutions? Which sorts of problems are such institutions and processes best suited to handling?

Berry

Berry—What Is Society?

For Wendell Berry, the idea of "society" is closely tied up with the concept of sociability, the cooperative relations one has with members of one's family and community. That there is a larger society of people one does not know personally and of states and nations to which one has a responsibility cannot be denied, yet Berry worries that such entities tempt us to take refuge in abstraction. He argues that

> one cannot live in the world; that is, one cannot become, in the easy, generalizing sense with which the phrase is commonly used, a "world citizen." There can be no such thing as a "global village." No matter how much one may love the world as a whole, one can live fully in it only by living responsibly in some small part of it (*UA:* 123).

To take responsibility for the whole world is to overreach ourselves. We are apt to substitute rhetoric for understanding (*D&H:* 89) or to take on roles that detach us from responsible connections with the people and places nearby and from some parts of ourselves. For, just as there is a limit to how much land one can farm, there is a limit to how large a society one can responsibly interact with.

The two most important social units for Berry are the family (and particularly the spousal relationship) and the community (*HE:* 113-8). The spousal relationship is especially important for Berry because it is the human relationship that embodies fertility and reproduction, and thereby, cyclicity and permanence. In "The Body and the Earth," the central chapter of *The Unsettling of America,* he argues that there are fundamental links between the way we view and treat the land and the way we regard and employ our own reproductive power. The links are not simply analogical. By exercising sexual and reproductive responsibility we are taking direct action to live responsibly within the bounds of biological sustainability. We are living within limits by not requiring a place to sustain us limitlessly. Berry is certain . . . that no satisfactory solution [to contemporary agricultural problems] can come from considering marriage alone or agriculture alone. These are our basic connections to each other and to the earth, and they tend to relate analogically and to be reciprocally defining: our demands upon the earth are determined by our ways of living with one another; our regard for one another is brought to light in our ways of using the earth (*UA*: 131).

With regard both to our treatment of the land and our treatment of one another we have sought technological solutions for social and cultural problems: "For the care or control of fertility, both that of the earth and that of our bodies, we have allowed a technology of chemicals and devices to replace entirely the cultural means of ceremonial forms, disciplines, and restraints" (*UA*: 133). Yet cultures exist which have humanely kept their populations stable for long periods through ritual, custom, and concepts of duty. Berry writes of the Hunza of northern Pakistan: "They had neither our statistical expertise nor our doom-prophets of population growth; it just happened that, placed geographically as they were, they lived always in sight of their agricultural or ecological limits, and they made a competent response" (*UA*: 133).

As important as the spousal relationship is the community, in two senses. First is the sense of a discipline of community that involves concern for the consequences of our actions—"community discipline imposes upon our personal behavior an ecological question: What is the effect, on our neighbors and on our place in the world, of what we do? It is aware that *all* behavior is social" (*D&H*: 156). Second is the sense of "community" as a group of people who interact regularly and directly in working out their livelihood together. Berry makes much of the New England farming community of the 1940s and 1950s—a community of diversified farms where much produce was sold locally and in which economic relationships were equally personal relationships (*GGL*: 100-1; *HE*: 179-92).

What Berry is sketching here is a notion of society in which the "big" problems have no chance to arise because the many small problems which constitute them are being dealt with by individuals—in marriages, families, and communities where responsible living is going on. For example, widespread practice of responsible reproduction (and agricultural production) would alleviate both social tensions and stresses on environmental systems. There would no longer be a need to exploit the land or exploit one another to feed a rapidly growing population of displaced and unplaced people. Many commentators on agriculture might agree, but would be less sanguine about the likelihood of humans learning or relearning how to live responsibly. From their viewpoint it is our inevitable irresponsibility that requires large-scale institutional solutions to social problems. But for Berry such appeals to inevitable shortcomings in character are examples of the retreat into abstraction and facilitate further retreat into more abstraction. Such solutions will not solve cultural problems.

Berry—Who Are We?

A key area of disagreement between Berry and his critics (especially former Secretary of Agriculture Earl Butz) is the practicality, attractiveness, and legitimacy of the way of living Berry advocates. To Butz, Berry's back-to-the-landism represents loss of freedom, virtually a return to bonded serfdom. He sees Berry as a romantic, and suggests Berry can be satisfied running a small farm only because Berry is not a "real" farmer, but has the income and diversions of a tenured English professor to fall back on (Butz 1978). But is combining teaching with farming hypocritical? Not to Berry.

Their disagreement has to do with how one characterizes one's role(s) in society—i.e., how we fit in, or who we *are*—and how one ought to think about the several social roles one plays (see Figure 10.1). Butz thinks farming is or ought to be the primary identity of farmers; people are defined by, and need to organize their lives around their primary economic function. Other activities, work or play, are for after the farming has been done and are secondary to one's vocation. An advantage of mechanization and monoculture is that they provide the means to farm enough land so that one can make a living and even acquire the means and free time for recreation.

Figure 10.1 Who Are We?

Berry	multifaceted persons	spouse & farmer & parent & neighbor & artist & doer
Walters	interests in conflict	farmers and consumers vs exploiters
Agriculture 2000	participants in the economy	farmer or agribusiness person or researcher or consumer or rural resident

Berry rejects the idea that the purpose of work is to give one the freedom to recreate and enjoy life. To separate work from "real living" demeans both: work becomes something we hate but must do, while our free time is spent in frivolity. Instead, we must reorganize our lives to reintegrate work and recreation. To be, like Berry, a farmer-professor-poet-novelist-essayist-husband-father-neighbor-etc. is nothing to apologize for, but instead an example of integrated and responsible living. To farm only "part time" (the concept is in fact alien to Berry's notion of an integrated life) does not mean one is a dilettante; to have an outside income does not mean one's farming is unsuccessful or uneconomical. Further, if we would accept the "technology of enough" in place of the "technology of the ever-more" we would find that responsible small-scale farming need not require constant labor (*GGL:* 169).

What we must do then is to stop seeing ourselves exclusively in terms of one (economic) role we perform in society, which at best artificially isolates one portion of our activity. Instead, we must regard ourselves in terms of all the things we do and bring to our work, for example, the same joy and commitment to quality we reserve for our "private" lives.

Berry—Who Are Others?

More than any of the other writers, Berry emphasizes the common elements of the human experience. The literary classics are classic, he argues, because they address perpetual problems of the human condition. And the possibility of responsible agriculture lies in a panhuman sensibility not limited to any single culture or class. Yet even in his works there are "others": the researchers and pundits who envision an agriculture with fewer farmers and even more automation, and those ostensibly sympathetic to farmers, like a weather forecaster on a rural radio station, who seem dangerously out of touch with farm living. Berry's emphasis on human agency, human potential, and human rather than institutional responsibility make the issue of how these people become "others" especially important. Unlike the other writers he is not in a position to explain their otherness in terms of institutional demands, class interests, or the existence of evil; too much of his vision is tied to a strong and positive view of human character to permit these explanations.

"Others" are to be thought of not as villains, nor dolts, nor markets to be satisfied, but as individuals much like ourselves, though more confused, less aware, further alienated from viable culture. Specialized educational institutions have blinded some people so that they can no longer perceive implications and connections, especially the connection between what we do and what happens to others and the environment.

This issue is especially acute with regard to agricultural researchers who grew up on family farms but are now contributing to the disappearance of their heritage. He is sure that they must have doubts, "some of them may occasionally overhear their critics with a tremor of recognition; from time to time some of them may even come face to face with bad external results of internal purposes and recognize them as such." Yet they cling to "self-protective orthodoxy . . . out of fear of knowledge of another kind. . . . [they are] *desperate* to define the *only* possibility" (*UA*: 172). Here too the failure is a human failure; a combination of fear of the unknown and the need for the security of stable authority accounts for the closed-mindedness of agricultural scientists. Berry does not deny that the institutions of research may increase this defensiveness, but neither does he go out of his way to point that out. We are left with a picture of persons who have experienced the illusory security of specialization and who now need to find courage to face the complexity of the whole.

Elsewhere, in an essay "Whose Head is the Farmer Using? Whose Head is Using the Farmer?" Berry criticizes agricultural journalists and radio experts who broadcast advice to farmers. Their glib confidence suggests "that farming is a business in which there are no real problems." Such messages can have no credibility for farmers; the "world of radio agriculture" is simply "not the same world that farmers live in" (Berry 1984: 22-3). Unlike Walters, Berry sees nothing particularly sinister in this banality; he wonders only how it is that real people can honestly concoct it: "these people [agricultural journalists and experts] undoubtedly have lives, at-home lives, that have certain things in common with the lives of farmers: uncertainty, worry, bad surprises, loss of confidence, fear of failure, failure." Here the appeal is to commonality. These people have lost touch with the lives of their audience; the fantasy of the city has disconnected them from the problems of farmers' lives.

What is noteworthy about these passages is that Berry takes care to avoid explanations of otherness that rely either on dehumanization—the image of others as fundamentally different from us—or determinism—the image of society in which people are simply acting out roles that they do not define. Instead, he lays out a common ground of humanness, a set of abilities and anxieties presumed to be familiar to all. From these he explains both what goes wrong (how we become problems for ourselves) and how we can resolve those problems (see Figure 10.2).

Figure 10.2   Who Are They?

Berry	people like us, only	– out of touch, more confused – unaware of consequences, implications, connections – more fearful, insecure – less open-minded
Walters	exploiters	– big business, e.g., grain traders – high finance, e.g., bankers – government officials, academics, clergy
Agriculture 2000	different groups in different contexts	– farmers (to input suppliers) – consumers (to food marketers) – suppliers and marketers (to farmers)

Berry—Roots of Social Problems

In Wendell Berry's view social problems come mainly from the alienation of people from the land and from meaningful ways of living. The sources of the problems are thus in the institutions and practices that alienate, and the solutions come as individuals change their ways of living. Berry's perspective is less apparent in the problems he describes than in his understanding of their origins and of what their solutions must be. Consider for example his description of the problem of the welfare of the poor and ill:

> We now have millions on some kind of government support, grown useless and helpless, while our country becomes unhealthy and ugly for want of human work and care. And we have additional millions not on welfare who have grown almost equally useless and helpless for want of health. How much potentially useful energy do we now have stored in human belly fat? And what is it costing us, not only in medical bills, but in money spent on diets, drugs, and exercise machines? (GGL: 132)

Some might read such a passage as an attack on public welfare, others might view it as a description of a health problem to be dealt with by public means, but for Berry it is a commentary on the degree to which Americans have lost the habit of taking care of themselves and one another. What looks like a social problem is in Berry's view a collection of private and personal problems, the solutions to which must ultimately be discovered and implemented by the individuals themselves. This theme takes the form of a programmatic statement in the essay on "The Reactor and the Garden" discussed earlier. There Berry argues that complaining about nuclear power or picketing nuclear plants does not solve any problem: "In our efforts to correct the way things are, we are almost always, almost inevitably, opposing what is wrong with ourselves. If we do not see that, then I think we won't find any of the solutions we are looking for." The problem is not the risk of the power plant but our dependence on the energy it produces, which we use to escape doing what we can and ought to do for ourselves. The solutions come from reducing dependence, for example, by growing some of our food in a backyard garden.

Berry—The Role of Government

Somewhat surprisingly, neither Berry nor any of the other authors is preoccupied with agricultural policy. As we noted in Chapter 7, Berry is uncomfortable with the idea that government programs can solve problems located within individual lives: "free men are not set free by their government. Free men have set their government free of themselves; they have made it unnecessary" (*D&H*: 129). The appropriate role for government is therefore a negative one: it should "protect the small and weak from the great and powerful," but not become "the profligate, ineffectual parent of the small and weak after it has permitted the great and powerful to make them helpless" (*UA*: 219). Governments may clear the decks of impediments so that free people can live responsible lives but they cannot deliver freedom or legislate responsible living.

This view comes across clearly in Berry's opposition to a TVA dam-building project. The dam would create a lake for recreational use but would submerge the homesteads of 949 families. For Berry this is an example of "the tyranny of 'public service'—the homes of 'the few' high-mindedly sacrificed to the 'recreation' of 'the many'" (*GGL*: 80-1). He has two objections to the proposal. The first is that an abstraction, "public good," has primacy over the concrete and particular: the responsible lives of real people living in the particular place of their settlement. "Homes" were to be submerged and a "home" for Berry is not just any old place where one happens to live, but the site of sacred commitments and relationships involving family, land, and community.

The second objection is to the sort of recreation the lake will provide. It is recreation within set times, according to set rules, available only to a particular segment of the population. Moreover, it embodies an absolute distinction between recreation as what we do in our leisure, and work as what we must do to live, a distinction that trivializes recreation and dehumanizes work. Elsewhere Berry objects to this kind of artificial and delivered freedom: "People will be *allowed* to be free to do *certain* things in *certain* places prescribed by *other* people" (*UA:* 74). In his own ordering recreation takes on a different cast: it is not distinct from work but should be part of work; it is not trivial or peripheral but essential for health. For Berry recreation is akin to re-creation, a transformation simultaneously physical, mental, and spiritual. Far from representing an escape it is centered in home or community as part of a responsible and settled life.

Yet one might still argue that the displacement of persons living responsibly on the land is warranted because public benefits outweigh the losses. To nearby city dwellers with no opportunity to develop responsible relationships with farms, this project will be the only means of recreating in nature. Berry's rejection of this argument is more than a rejection of its utilitarianism; he is concerned that such a policy misrepresents the kinds of problems public action can solve. For him, the perception that there is a need for a recreational escape for city dwellers signifies only that city dwellers (impeded perhaps by public policies) have not found ways to incorporate recreation and responsibility into their own lives.[1]

This view that what others will see as public or social problems are really problems of individual lives is also apparent in Berry's discussion of poverty and hunger in third-world countries. To many writers, world hunger is quintessentially a public problem, a problem of gathering the requisite proteins, calories, vitamins where they are abundant, and of getting them to where they are needed. Berry takes a different view. In a 1979 essay "An Agricultural Journey to Peru" he contrasts slum life in Lima with life in the Andes, where strong culture and sustainable agriculture survive. In Lima are those who have left marginal livings in the country for marginal livings in the city, where they find inadequate housing, employment, sanitation, and food. Bad government land policies may have led to such conditions Berry recognizes, but a solution will only come through culture as people learn to live within limits, as upland Andean farmers do. He is mainly concerned not with policy (or with demography; overpopulation, he admits, is a great part of the problem), but with human dignity—with indications (e.g., flowers in a door yard) that people retain a capacity for responsible intervention in their surroundings and, therefore, a residue of culture (*GGL:* 6).

Though more concerned with the renewal of culture, Berry does recommend some specific policies, but only after reiterating his wariness of "big solutions": we need policies to make farmland available to people who want to farm, price controls that benefit operators of small farms and farmers who do not over use their land, and programs that encourage local self-sufficiency and recycling. We also need to eliminate excessive sanitary regulations that discriminate against smaller operations and to encourage diversity in plant and animal stocks (*UA: 218-22*; HE: *123-32*). Yet for any of these to be effective there must also be changes in the sensibility of individuals towards themselves, others, and the land.

Walters

Walters—What Is Society?

In Berry's view all societies face the same problems of the human condition. Regardless of where we live, the problems we encounter—in living on the land, in marriages, and in communities—will be similar, though we will seek to solve them in different ways. From Walters' perspective this view is naive, for it ignores fundamental conflicts of class or group interests, such as between farmers and food processors. It will be more important for farmers to be effective lobbyists than to have learned to live within cultural and ecological boundaries. Family and community will be less important than regional, national, and international structures for collective bargaining and for setting economic and environmental policy. While conflict of interest must exist, it need not be violent, nor need it interfere with the pursuit of efficiency, health, and ecological sustainability, provided it takes place within what Walters sees as the true laws of social and economic organization. Just as there is a transcendent Nature whose laws can be used to liberate us and improve conditions, so too there are transcendent social laws that we must abide by if we are to bring about what is just and fair and, therefore, stable and enduring. As long as institutions are consistent with those laws, all will be well; conflict will occur, but within a stable equilibrium. But what has happened in the U.S. since the end of the second world war is that one side in the conflict—the "tight pants boys on Madison Avenue and Wacker Drive" who conspire to control farm prices (*AT: 65*)—has become so dominant that institutions no longer embody the laws of economic and social balance; American society has become dangerously destablized and is on the brink of economic disaster.

The laws of a just society which most concern Walters are laws of production. His most detailed discussion is *Unforgiven, the Biography of an Idea* (1971). The book focuses on Carl Wilken, an Iowa farmer and self-taught economist, and the Raw Materials National Council, a private New Deal-era organization headed by Wilken, University of Wisconsin agricultural economist John Lee Coulter, and Charles B. Ray, an economist and executive with Sears Roebuck (*U:* 115-78, 190-91). Their argument recalls the eighteenth century physiocratic view that since what the earth gave was free, agriculture and other primary industries, like mining, were the source of all new wealth. (In fact Wilken relied more on the atomic scientist turned economic theorist Frederick Soddy, who held that wealth could be measured as released energy—but again it was only primary production that created wealth (*U:* 45; *AT:* 136-40).) Wilken and Walters agree that the surplus from primary production will ramify throughout the economy so that each farmer's profits will support seven non-farmers through the farmer's purchases of farm inputs and consumer goods and services. By contrast, the wealth of Keynesianism was merely fictional wealth that was ultimately driving the only real-wealth producers out of business.

Hence to Walters, the "adaptive" approach of conventional agricultural economists who favor low food prices and see as necessary the movement of farmers out of agriculture is entirely wrongheaded: there cannot be real economic growth if there is no source of new wealth. On the other hand, higher prices for the farmer's produce will produce real growth since the more profit the primary producer makes the more money will ramify through the economy creating wealth for all. In some respects, this view is antithetical to Berry's. Where Berry is concerned about undisciplined release of productive energy, Walters believes we must maximize primary production (and consumption) and unleash energy since this is the only way to create wealth: "The American economic problem is to consume more, not less, food, and to consume it at full parity prices" (*AT:* 140). It follows that more frugal ways of living are not desirable; for example, we cannot have a civilization based on vegetarianism because it produces too little wealth:

> Instead of driving the economy toward soybean mushes, artificial milk, and low feed grain conversion meats such as chicken, the policy makers would do well to move toward a rare meat that converts 14 pounds of grain into a single pound of meat. . . . The low feed grain-meat conversion figure—an advantage at the individual level, anathema at the body economic level—contributes mightily toward upsetting the balance between production of consumable wealth and permanent wealth (*AT:* 140-1).

In Walters' view the desire for wealth is innate. We can choose, however, to satisfy that desire in socially desirable ways (e.g., by producing consumable goods or food) or in ways that are pathological, such as through colonialism or militarism (*AT:* 138-42). We must judge actions with regard to macroeconomic systems and adjust policies to the laws of just social and economic behavior.

Walters—Who Are We?

For Walters, who we are is a function of our class and occupation; there are no responsible ways of living that transcend these. The Amish, who to Berry demonstrate how people in a strong culture can retain control over their society even when the society around them is out of control, seem to Walters to be clinging to an illusion of independence. So long as they participate in the agricultural market and pay taxes, the Amish are not independent, and are being exploited as much as other operators of small and midsized farms. He notes that eventually even the Amish joined the milk holding action.

> They had accepted the economic consequences of . . . going it alone. They wanted no truck with the outside. They wanted only to do their own work, rear their children and care for their old. They had accepted low income as a status symbol, and willingly created their own furniture and homes with their bare hands. But even the Amish realized they were being crowded. They were being crowded by milk regulations, taxes on their land, burdens from civilization that meant they had to exchange some of the wealth they produced for wealth created in other segments of the society. Since they were not being paid for as much wealth as they produced, these exchanges served only to rob them . . . of the fruits of their toil (*HA*: 212).

These conflicting views of the Amish reflect two contrasting kinds of solidarity, two ways of thinking about who we are and what we have in common with others. For Berry, Amish solidarity reflects a great deal more than the fact that they are people who have chosen to live outside contemporary society. Through their common faith they have a commitment to one another, their families, and the land they farm. Since all people can make similar commitments there is potentially a panhuman solidarity that can and should be more important than identifications of race, creed, class, or nationality. For Walters, what unites the Amish as a community is their common strategy of providing for themselves services that would otherwise be provided by the public. But because they can never completely fence themselves off from the chaos outside, their solidarity is false and dangerous; it is the perfect internal solidarity of a union local unwilling to commit itself to the solidarity of the labor movement as a whole.

Thus for Walters, how we fit in is dictated by our situation with respect to the rest of society, the way the rest of society treats us. We achieve liberation or autonomy only by recognizing that, and confronting and challenging it, not by pretending it doesn't exist. The clearest illustration of this is Walters' description of the rugged individualist who could not dump milk, considered already in Chapter 7.

> Allowing milk to slide across the floor to the nearest drain was to them like dying for a cause alone and unheard of, an absurdity no consideration in heaven or hell could make real. The overview of the rugged individualist ranged no further than the farm, this crop, that cooler full of milk, and all else "out there" was too complicated to grapple with. The individualist could only fall back on God and nature—and prayer—if his stuff wasn't fetching what it was worth. Like the die-hard in any pact with futility, he would suffocate never realizing he was being choked to death by the marketing system. He would look on those who banked at least a little hope on collectivity as less than "he men" because they refused to let themselves be chewed up and spit out by the powers that be (*HA:* 171).

Berry would be uncomfortable being cast as Walters' "rugged individualist," but the passage does raise some of their differences. Berry writes in "Discipline and Hope" that the "good farmer" plants without regard for price or market, but "in response to his discipline and to his place" (*D&H:* 138). For Berry understanding one's smallness in the universe is necessary, and God and prayer are not refuges of fools but ways of coming to grips with the world. In his view one escapes futility by living in a way that keeps ecological and cultural systems intact. One does not have to take heroic action to escape futility; one simply needs to find purpose within the bounds of ordinary life.

Walters—Who Are Others?

Preoccupied with class conflict, Walters sees a sharp distinction between "us" and "them." His rival classes follow the traditional populist division of society: a dominant "them"—big business, high finance, government and academia, who are corrupt and predatory—against an exploited "us"—farmers and workers, land owners and land renters, *petit bourgeois* and peasants, men and women (*HA:* 139). Class is thus not a neutral concept for Walters: classes are not empirical constructs for subdividing a population but opposing parties in the struggle for justice. Placing people or occupations in a class thus involves a moral evaluation: are they perpetrators of injustice or its victims?

In this dichotomy "others" are dehumanized. By contrast with Berry's painstaking attempt to find a common humanness even in those whose views or actions he deplores, Walters is contemptuous toward the opposition: bankers and grain traders are "the tight pants boys on Madison Avenue and Wacker Drive" (*AT:* 65) or "the bright boys in the carpeted offices" (*AT:* 15). Academics and clergy are "humpbacked under the weight of academic degrees, . . . auto-hypnotized by their own voices" (AT: 291). Government officials and others who oppose NFO hog kills yet force farmers to take extreme actions are "sob brothers" or "cry reps" (AT: 221, 292).

These people are presented as having lost what humanity they once had. But how do they get that way? Are they brainwashed by ideology? sincerely misguided? or just evil people? Most of them, Walters is convinced, are simply too dull to see what is wrong with the existing order. Farmers don't see the necessity of collective bargaining and farm boys are easily co-opted into business, politics, or academia (AT: 212, 269, 273). As we saw with Staley, accurate insight is a prerequisite for righteousness; we must have the acuity to recognize justice and truth if we are to strive for it. But there are equally aware people on the other side; the evil geniuses who control the nation's wealth are portrayed as fully aware of what they are doing (AT: 277).

There are thus four groups in Walters' scheme, two on the "us" side, and two on the "them" side (Figure 10.3): (1) those like Staley (and Walters and the readers Walters addresses) who can be presumed to be virtuous and aware; (2) the mass of common people who are victims, but who remain like "sheep," too unaware to assert themselves (AT: 273); (3) those in exploitative institutions who do the will of their masters under the mistaken idea that they are doing good; and (4) the evil geniuses who hold the power and "call the shots" (AT: 277, 280). Of these four groups, there is hope for the second and third if they can be made aware: Walters notes that Carl Wilken addressed his theories of economics to the "man on the street, the same one who stood guard over the fact that two and two were still four" (U: 55). The mass of humans still retain enough wits and decency that great social changes would be possible if only people could be made to understand their necessity. The fourth group may be unredeemable: they are "minds out of control," utterly and irrevocably different from and dangerous to the rest of us (AT: 277).

Figure 10.3 Walters' Social Groups

		Victims	Exploiters
	Aware	prophets and leaders (Staley)	evil geniuses
	Unaware	most farmers and common people	duped pawns

Walters—The Roots of Social Problems

For Walters, social problems begin when we fail to follow rules for living which are evident in nature and enforced by natural laws of causation. These rules, as we have seen, relate both to our biological existence—we must have sound food to thrive—and to our social and economic affairs—we must take care that primary producers prosper. Our failure to heed these rules manifests itself continually and pervasively. Indeed, we are now in a dangerous situation in which our economic and educational institutions systematically violate these laws, and the hope for change rests therefore on thoroughgoing institutional reform. The rewards of such reform will be far-reaching. In contrast with Berry, who sees a need for each individual to learn the difficult skills of responsible living, Walters holds out hope that once our institutions incorporate nature's laws our great social problems will have been solved. Individually, we will be better human beings—longer-lived, healthier, more mentally acute, and hence better able to look after ourselves and to act responsibly. Conflict and competition will still exist but within a just balance of power. Because the rights and important roles of primary producers will have been recognized, public policy will protect the interests of farmers.

Walters—The Role of Government

Of the three authors, Walters sees the most room for effective government action. Yet he is hostile toward the policies of the past and present, and skeptical of future policy.

The main way governments can contribute to the welfare of farmers is by establishing full parity prices for farm products. During the heyday of the NFO, Walters held out hope that fair pricing could be obtained independently of governmental action through contracts between the NFO and processors. But it was government action—an injunction obtained under the Sherman Antitrust Act against the NFO's 1967 milk holding action—that brought down the movement. More recently Walters has argued that parity is essential in the reformation of agriculture (*HA*: 332-6; *CEA*: 20-38; *Acres* 1/82: 20-1, 2/82: 2).

Walters argues that the attempts of governments to rectify farm problems have been ineffective because they have been subverted by the dominant interests in society, who exploit farmers. In the nineteenth century "unprincipled men" found they could "draw power from the second raters in the society simply by *delivering*" (*AT*: 39). They provided the social welfare programs that dissipated farmers' anger and bound people more tightly to the institutions that are exploiting them. Low-income housing and similar programs have "institutionalized poverty." In making it easier for poor people to survive without confronting their exploiters, these programs have obscured the need for radical social change (*HA:* 74). Likewise, hospitals and sophisticated medicines have kept bodies patched together and diverted attention from the poisoning of humans and the earth (*HA:* 179). The dominant classes pretend to serve the needs of farmers in formulating agricultural policy, but the price support programs they establish do not give farmers the security they need to prosper. The sharp-minded and self-reliant settlers of the country have thus been transformed into a population of obedient sheep (*AT:* 273).

While Walters sees in public policy the power truly to solve problems (as well as to deceive people into thinking they are solved), he holds out little hope that the current system of government will solve them. This is not because the problems are unusually complicated: quite the contrary, the laws of production are simple. Effective government action is unlikely because the masses of people who would need to make themselves heard are too dull, tired, and easily distracted. So pessimistic a prospect might seem to undermine the possibilities for social change completely, for Walters seems to see no chance of action in the only forum in which effective action can occur. Yet the prospect is not so grim. We explore Walters' more hopeful views of the bases of effective social action in the next chapter on Praxis.

Battelle

Battelle—What Is Society?

Like Berry and Walters, the Battelle report finds a particular stationary moral and social reference point, a high ground from which the rest of society is viewed and evaluated. In *Agriculture 2000* that fixed point is "the economy." The economy is not a set of disembodied laws as it is for Walters, but the totality of economic participation, institutions, and goals. The condition of the economy represents the condition of society. It is in its best state when it is growing and bringing about a higher standard of living, and our central task is to manage things so the economy does grow. This conventional view is assumed as a given in *Agriculture 2000*.

It may seem strange that so abstract an entity can function as a reference point; one wants to ask *who* is the economy and *what* are its interests? We answer this question below; here it is enough to note that "economic" and "social" are sometimes used interchangeably. Both are more than collective terms: one may speak of both economic and social "forces" or "norms," and in several ways one may ascribe to the economy functions that might be conceived as social. For example, the economy can supply an identity—one might define oneself and others in terms of economic roles. It can also supply social boundaries—one might distinguish normal participants in the market (members in good standing) from those who reject or pervert the market. The concept can also supply norms and notions of how the world and society work: one can appeal, for example, to "economic rationality" to judge actions, to discover both whether they are morally good and whether they are likely to produce the ends sought. Finally the economy can provide purposes and visions: one's vision of a better life—which can be conceived as the economist's "rational self-interest"—can shape how one spends one's days, why one takes on hurdles, why one is satisfied (or dissatisfied) with what is accomplished (see Figure 10.4).

Battelle—Who Are We?

For both Berry and Walters the crucial "we" is, at least in a loose sense, farmers. For Berry there is potentially a "good farmer" in each of us; Walters speaks directly to commercial farmers. But in *Agriculture 2000* there is no single reference group. In the society-as-the-economy motif predominant in the report, each of the parties participating in the market has a claim to the privileged position of "us." The narrative frequently switches point of view, identifying variously with farmers, consumers, marketers of farm supplies or farm produce, researchers, or administrators of agricultural policy. As it does so, whatever group is being identified with occupies the role of "us" and is treated as having the power to act autonomously in just and proper pursuit of its interests. Groups with which it is interacting or competing are then usually treated

Figure 10.4   What "the Economy" Tells Us

Who am I?	a worker, consumer, producer
Who are they?	workers, producers, consumers
Who are outsiders?	those who reject or pervert the market
What makes things go?	enlightened self-interest
Where will the future take us?	to prosperity and progress

as "them," as objects or others to whom or for whom things are done. There can be only one "us" at a time, but by switching point of view the report is able to show how people representing competing interests and acting as free agents can still comprise a coherent entity, the market.

Because these groups are often in conflict, it is easy to become confused about how to evaluate their activities. For example, just as we have finally gotten used to seeing the world from the perspective of farm suppliers, we may find ourselves asked to adopt an opposing viewpoint. It seems that we must either take the authors to be moral relativists who find no single proper basis for evaluating actions—or we must recognize how they can have a moral (and conceptual) reference point that transcends the perspectives of the groups involved in agriculture.

The stable reference point that embodies the perspectives of all the competing groups is "the market." Because the market is presumed to resolve conflicting interests in the best interest of all, it can serve as a normative reference point. Still, accepting this doesn't solve the problem of how we as individuals fit in. Even if we believe that what the market does is right, we still need some guide for our perceptions and actions, a way of anchoring our sense of purpose and propriety in some particular group. In *Agriculture 2000* "agribusiness" is the term that comes closest to combining the normative status of the market with the unique point of view of a particular group. It is used in two senses: in a narrower, traditional sense to refer to a particular group or sector of the economy and in a larger sense to refer to a rationalizing spirit that characterizes that group and which, in the view of the report, is taking hold throughout agriculture. These two senses are often conflated and it is their conflation that allows the report to jump back and forth between the outlooks of particular groups and yet maintain an image of a coherent social order.

Something of this conflation is evident in the statement that agribusiness is "the *control center* of the American food and fiber system" (Battelle 1983a: 71, ital ours). Here the dynamic and organizational aspect prevails; the metaphor suggests that U.S. agriculture is a controlled system with control located at one place, under one will, working to a common purpose. The sectors of the food industry that embody this "agribusiness" range from those supplying farm inputs to those concerned with the processing, distribution, and sale of farm products. That territory is expanding, and "as farming becomes more sophisticated, all these [agribusiness] firms will finance, develop, and commercialize various technologies. Farmers will increasingly rely on these businesses to provide the capital, products, information, and services vital for running tomorrow's farm operations" (71). In this passage "agribusiness" is seen as taking charge of what one might have called "farming": it provides services, e.g., of custom cultivation or harvesting, and increasingly, through development of the algorithms considered in Chapter 9, agribusiness consultants will supply professional guidance for farming decisions: "soil testing; tissue testing; individual field recommendations for plant food, pest management, and growth regulators; advice on livestock nutrition" (72).[2]

To include so great a range of activity under one term may seem to cover up conflict. Walters might object that farmers' legitimate interests are obscured when they and their activities are included in this nebulous "agribusiness." Yet from the viewpoint of *Agriculture 2000*, the breadth of "agribusiness" simply reflects the integration of modern agriculture. Whatever role one may occupy within this system, it is essential to understand that role in terms of its integration with other groups; to dismember "agribusiness" and go back to thinking in terms of independent groups would be to give in to sectorial squabbling and lose the benefit of coordinated thinking about production and distribution. And it is important that we think this way: whoever we are—farmers, food processors, seed sales people, even consumers—*Agriculture 2000* is concerned that our primary identification should be as members of an enterprise in which disparate activities are coordinated to produce a result that is stable, beneficial, and efficient.

Battelle—Who Are Others?

In *Agriculture 2000* no one group occupies the position of "us" throughout, yet the report often takes the view of "agribusiness" in its narrower sense. Immediately following the representation of "agribusiness" as a "control center" is a list of the "four big challenges [which] face marketers of farm supplies": (1) meeting the "needs" for products and services, (2) keeping up with "new technologies" and finding people who can transfer technology from labs to farms, (3) overcoming "farmer/rancher resistance to price (not just price increases)," and (4) meeting "the increasing costs of doing business" (71). One can see, from this list and from other passages, how the us-them division works, endowing one party with agency and legitimacy, seeing others as passive or as problematic.

It is first of all striking that the items on the list are "challenges" and not "problems," "issues," or "concerns." To take up a "challenge" is a voluntary action; only a coherent entity with moral, social, and cultural legitimacy can recognize and accept a challenge. The term also brings an expectation about the appropriate outcome: challenges are to be met successfully. These "challenges," then, are not just things agribusiness firms should try to do; it is right that they succeed in doing them. Thus to represent "agribusiness" as facing "challenges" endows it with unity, agency, and moral legitimacy. It also withdraws agency from those sectors that manifest the challenges; since the challenges are to be met successfully, they are put in the position of yielding; they become a "them"—objects to be acted upon.

In the third of the "challenges" to agribusiness we can see especially clearly how representing conflict as "challenge" reinforces the us-them distinction. Here agribusiness is challenged to "overcome" the "resistance" of farmers and ranchers "to price (not just price increases)" (71). It may appear that there is agency on both sides here—farmers and ranchers resist, agribusiness seeks to overcome resistance. Yet moral legitimacy rests wholly with agribusiness. In treating resistance as a challenge to be overcome, *Agriculture 2000* implies that it is pathological; it is something farmers and ranchers ought to get over. The parenthetical statement that farmers and ranchers resist price altogether, not just price increases, further suggests that the resistance is irrational—farmers and ranchers evidently want something for nothing.

The report also identifies the causes of farmers' resistance as abnormally low farm prices and farmers' inefficiency in marketing: "most farmers make decisions on selling crops the way they have for the last 50 years. Since that type of marketing is not very efficient, they try to increase profits by buying inputs as inexpensively as possible" (74). It warns that pressures on farm supply firms are not lessening, "because farmers want to buy at the lowest possible price (but still receive services)" and adds that resistance to price will grow "as the general level of farm managerial expertise rises" (74). It describes a practice common among farmers of obtaining bids for a complete package of farm inputs—seeds, fertilizers, pesticides—but insisting on "individual prices for each item" and then returning "to each supplier to buy only those items at the lowest price quoted . . . [among] all three bidders" (74).

Thus from the viewpoint of agricultural input suppliers, the challenge has several components: chronic low farm prices, farmers' "managerial expertise," and their aggressiveness in shopping around. Implicit in this depiction is the solution: farmers should pay what suppliers charge and acquiesce to a system of bidding for a full line of goods and services rather than bargaining separately for each. Notably, the problem here exists only when the situation is viewed from the perspective of farm suppliers; others might see only normal market operation. The prices farmers receive may be "realistic" prices, the outcome of supply, demand, and farm politics. Farmers' resistance to price might be seen as normal consumer behavior and their ability to minimize costs might be deemed praiseworthy. One could as readily suggest that it is the suppliers, unable to make a profit under such conditions, who are the problem. Yet so long as one tries to see the market from the perspective of a particular participant in it, as *Agriculture 2000* does, this view does not arise; in this section farmers remain the others and constitute a problem for the "us" (see Figure 10.5).

In some cases the report does endow both parties in a transaction with legitimacy and agency. In "Marketing Cooperatives" farmers are seen as joining co-ops to get better prices. Farmers' co-ops are advised that the best way to get higher prices is to market "a branded [i.e., differentiated] product." But here the discussion brings in consumers' views too. Co-ops are advised to "produce foods for specific consumer demands. The cooperative, through market research, determines what these demands are and provides farmers-members with crop and livestock production programs tailored to meet these demands" (65). Here farmers restructure their marketing to meet consumer demands directly, and consumers know what they want and tell farmers in market research surveys. The relationship is harmonious and symbiotic: neither party is exploiter or exploited, and the us-them dichotomy fades.

Yet as a section on "Supply Management" illustrates, discussions that appear to endow both sides with agency may be highly ambiguous. The section is concerned with production contracts between farmers and commodities buyers. According to the report, as agricultural operations become more systematic, predictions of yield will become more accurate, and it will be more attractive for buyers to contract to purchase a farmer's crop at an agreed-upon price. Such contracts will have advantages for farmers and buyers. The farmer will be able "to make a pricing decision" at some time other than harvest and presumably will not be pressured to accept a depressed price. Farmers will have shifted risk to buyers and gained better market access. Buyers will be able to guarantee quality and schedule deliveries to keep processing plants running at optimal levels.

Figure 10.5   The Rationalizing Spirit of the Market (Agribusiness)
             Mediates Conflict

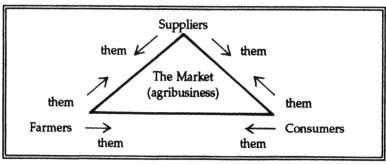

Contracts like these represent the rationalizing force of agribusiness ("agribusiness" in its larger sense) but because they change customary relations between groups, the groups no longer retain their previous identities. With the shifting of risk, the farmer ceases to be a small-scale capitalist and becomes a contractor providing services for an agreed price. The commodities buyer becomes less a gambler and more a manager, concerned with quality control and utilization of plant capacity. With processors able to keep plants operating efficiently, and farmers assured a stable income, conflict declines, and the need for intragroup solidarity fades, taking with it the us-them opposition.

But whether such contractual relations reflect the activity of parties with equal legitimacy and agency, and can be the means for transcending longstanding conflict, is open to debate. Representing an ideology in which class conflicts such as those between farmers and agribusiness are inherent, Walters sees such contracts quite differently. As NFO publicist, he championed a contract production scheme much like that advocated in *Agriculture 2000*: under the auspices of the NFO, the prices farmers would receive would be negotiated with buyers well before harvest. Yet for Walters and the NFO, the key issue was not the contract itself, but the terms of the contract, an issue on which *Agriculture 2000* is silent. For Walters, contracts were to be evaluated as just—embodying a price sufficient to maintain the livelihood of independent farmers—or unjust—embodying a "take it or leave it" price set by buyers. Contracts with unjust prices would represent an unequal and exploitative relationship, the subjugation of farmers by buyers. Chicken production in the south exemplified such unjust contracts to Walters. There farmers had become virtually indentured to large corporations.

Such differing assessments of the significance of contracts reflect very different conceptions of social structure and of the origins of social conflict. For *Agriculture 2000*, conflicts between farmers and buyers are transitory and aberrant. Their prominence in the past has been due to an inability to control production, due to the vicissitudes of weather or pests, both of which we can increasingly recognize and correct for. Farmers and buyers may always have competing interests, but sooner or later, through the adaptive forces of the market, supply and demand will come into harmony: the common goals of efficient production and economic success will eliminate overt conflict. As the rationalizing spirit of "agribusiness" subsumes both farmers and buyers, members of these groups will cease to see themselves as separated by distinct interests, social structure, or visions and values. "Us" vs "them" will no longer apply; all will be "us," engaged in a common enterprise. For Walters, on the other hand, conflict between farmers and middlemen is inherent, a concomitant of true economic freedom. Contracts may suppress or stabilize conflict temporarily, but they will not eliminate it.

Battelle—The Roots of Social Problems

Like Walters, *Agriculture 2000* sees social problems as arising out of the struggle to extract a living from nature, but unlike Walters it holds out no possibility of a golden age when we will have eliminated social problems. Because the report rejects the concept of nature as a purposeful entity, it has no room for natural laws that promise justice and order. Instead, humans are seen as continually under stress from nature. As our population grows, we need more space, food, and energy. Because nature does not supply resources as readily as we need them, we are left with disparities in wealth, health, opportunity, or leisure—the sources of our most serious social problems. We solve these problems as they occur as well as we can, but it is unlikely that we will solve them once and for all, for we will continually need to find new ways to alter nature to solve the problems our expansion is creating.

There may however be plateaus of accomplishment, periods during which particular problems that have long vexed us remain unproblematic. An example is the contract agriculture discussed above. *Agriculture 2000* portrays the coming of production contracts between producers and processors as a response to a new degree of precision in agricultural production. This new precision will permit us to limit production to demand and will presumably end the social and political problems that arose from years of surpluses, when prices fell and farmers were threatened with bankruptcy, or from years of famine, when consumers starved or rioted against high prices. In this way new knowledge will have ended one of the longstanding problems of society, the conflict between farm and city, and we can hope for other discoveries and developments that may similarly lay to rest others of the social problems that plague us.

Battelle—The Role of Government

Despite the conspicuous role of the federal government in U.S. agriculture, *Agriculture 2000* does not address the issue of future agricultural policy directly. The longer in-house report from which it was drawn does include a chapter on "The Government's Role in United States Agriculture," but that chapter deals with the history of agricultural policy rather than future policy needs.

But even if the report does not deal directly with policy, a policy agenda is in it. *Agriculture 2000* has spelled out technical and structural changes that will occur; the appropriate public policy will be that which keeps the food and fiber industry up-to-date with those changes. Unlike Walters, for whom policy is a means (if an impracticable one) of reforming an unjust society, the Battelle report sees policy basically as reactive, as reacting to the way things are and changing as they change. There may be periods of poor fit between policy and conditions, and a lag in the response of policymakers, but the key problem is not to change the basic thrust of policy, but to adapt it, to update policy to respond to the exigencies of the present and the needs of the future. The central concerns of the report are with policies that expedite development of new technology. It addresses such policy problems solely from the viewpoint of the advocates of new technologies, takes their view to be universal, and sees the policy problem as one of overcoming obstacles. All this is in keeping with the us-them perspective considered earlier.

Consider for example such statements as "successful development [of robot farming] will hinge upon the availability of research funding" (133). Since the authors have already suggested that such technology is the way of the future, they are indicating the role that public policy must play in securing that future. The same point arises with respect to development and commercialization of brassinosteroids, a growth regulator. The authors note that

> brassinosteroids offer the potential of increasing crop yields at little additional cost. Application is relatively simple, and may be achieved using existing equipment. . . . Commercial use also will depend upon approval of the compound by the U.S. Environmental Protection Agency. The environmental impact of brassinosteroids on insects, wildlife, and man must be assessed (102).

Here policy considerations are an afterthought; to mention a regulatory hurdle after the benefits have been spelled out suggests that the outcome of regulatory investigation is not in doubt, but that in the current policy environment good ideas must jump such hurdles. For Berry or Walters, by contrast, questions of the necessity of the new technology and its environmental (and social) impacts might well be the primary questions. Only when these had been answered could we go on to think about reaping the benefits.

The Battelle authors see the social issues raised with respect to displaced laborers in a similar way, as an "obstacle" to progress:

> An *obstacle* to the development of these technologies is *resistance* from people in the farm labor unions and from other people concerned about the effect of mechanization on the rural way of life. Concern has also been indicated about using state and federal funds to develop labor-saving devices. Such policies reflect the conflict that often arises between innovative technology and potential job displacement (128, ital ours).

Terms like "obstacle" and "resistance" make it clear that these concerns, however understandable, are undesirable to the degree they delay the benefits future technology will give us. We are warned to reconsider before making policies that impede our pursuit of efficiency. It is clear in such passages that the policy issue is being considered from one point of view only; the appropriate outcome of policy deliberation—to advance technology as rapidly as possible—is not questioned.

Yet such passages also make clear that the obstacles to progress have themselves been the product of past policies and politics—e.g., policies for environmental impact assessment—that have been influenced by factors other than pressures to adapt to market reality. The report recognizes that politics with all its quirks goes far to define the conditions in which technological and economic development must occur; in democratic societies there may be no perfect markets, and any technology may have consequences that some people will object to. It also sees a practical need in many cases for political compromise.

The clearest example of this acceptance of political reality is the section on "Exports." The very title "Exports" presupposes an us-them division. In the international marketplace policies advantageous to one country may not be in the interest of others. Here the "us" is clearly U.S. agriculture, and in particular the large firms that can buy and sell in bulk, market "value-added" products, and survive extreme fluctuations in price. The report equates the interests of these firms with the interests of the United States: "Such trade would be good not only for all segments of agribusiness but also for taxpayers in general. The export of agricultural commodities can help the U.S. achieve a balance of payments" (67).

But the report also considers "challenges" to the expansion of U. S. agricultural exports: the strength of the dollar, the "cost of credit," "protectionism," competition from other countries which export commodities, and all manner of other monetary, economic, and political policies and conditions. It recognizes the great difference between export policies that would be optimal from the perspective of the U.S., and those likely to come out of negotiations among several legitimate parties.

Here *Agriculture 2000* has moved away from representing problems only as they are seen by one group. The sharp dualism between the challenged and the challenger is muted; here "challenges" are complex natural and geopolitical conditions, not the wilful errors of uncooperative others. Take, for example, the challenge of "protectionism": "Foreign governments may restrict market access of U.S. goods, denying U.S. farmers the privilege to compete in potentially profitable markets." Here *Agriculture 2000* treats protectionism as arising from the interplay of legitimate political interests, not the irrationality of an out-group. The foreign governments possess agency: they "may" (or may not, as they choose) impress duties on our goods. The term "privilege" implies that being allowed to do business in another country is not a normal and natural right, but a gift that countries may grant or withhold. Even in the section on "other competition," our competitors are seen not as enemies, muscling in on our corner of the international market, but as others who can compete because they too "have increased their efficiency in producing agricultural goods," pressing us to increase our own efficiency.

Here what might have been seen as "threats" to U.S. agribusiness blend into conditions of economic existence. What might otherwise be "obstacles," are figured into the totality of factors that affect production, distribution, and consumption. We may deplore these conditions, but they are not seen as pathological; as someone's fault, to be rapidly removed by somehow forcing foreign buyers to meet their obligation to buy from us.

Unifying the discussion of policy in *Agriculture 2000* is the notion of policy as a response to change. We make policy not so much to facilitate responsible living (Berry's view) or establish justice (Walters' view), but to keep up with market reality. Berry and Walters might envision an end of policymaking: once governments had established proper social conditions, they would have little to do. The Battelle report sees change and progress as unflagging: to keep current we must continually revise policy, taking into account the most up-to-date science and technology, utilizing our best and most creative negotiating skills.

Notes

1. Berry does suggest that responsible living in cities is possible (*D&H* 135).

2. As it comes to include what formerly would have been "farming," "agribusiness" will displace the older terms "farming" and "agriculture." The farmer becomes either a consumer of "value-added" products and services, or an historical artifact, like the Amish farmer. It may be noted that increasingly one hears the term "grower" to refer to the modern practitioner of conventional monoculture.

11

Praxis

The topic of praxis lies closest to the primary concerns of makers of agricultural policy, for praxis deals with how people act to make their visions of a better future come true. We focus here on two questions: first, on what sorts of futures our authors think we ought to be working to achieve; and second, on their conceptions of how the actions of ourselves and others and of groups, professions, and institutions can create that future. For these questions parallel treatment of the three authors is difficult. Berry and Walters are critics of contemporary agriculture and envision a future distinctly different from the present. Each in his own way believes ordinary people under suitable circumstances can make significant social changes. For *Agriculture 2000* the questions are less appropriate, for the future will not be radically different from the present. There will be more—more food, prosperity, health, and leisure—but no conscious deviation from the present path is contemplated. Achieving this future will not require us to step out of our ordinary routine and regard ourselves as makers of significant changes (Figure 11.1).

Figure 11.1   What Is the Most I Can Do?

Berry	live responsibly
Walters	live righteously
Agriculture 2000	live normally

Berry

Berry—The Future in Question

Wendell Berry advocates a new relationship between people and land. He envisions a fully settled landscape, in which families would have roots in a particular place, in which more people would be engaged in an agriculture based on kind and careful land use. We come to this more settled existence not by leaving the everyday lives we now lead but through living those lives more responsibly. This is the theme of "The Reactor and the Garden," considered before: though there may be occasion for protests or exclusively political acts, the roots of problems are in ways of living and it is there we solve them. Thus political praxis is not something over and above responsible living; it is a way of living, of making our living count in realizing our visions of a better world. We cannot wait till the time is ripe to act but must enact our visions as well as we can now.

To others this view is likely to seem too timid or too credulous. It may seem to accept the injustices and inefficiencies of the present or to exaggerate the effects of good character. It may seem but a retreat to the simple and familiar in the face of large and complicated problems of safe and efficient production and equitable distribution. This misrepresents Berry. He is not seeking an easy solution that seizes on small goods and ignores big evils, but arguing that in indulging our fantasies of progress and panacea we neglect the difficult problem of living responsibly in the present. We find ourselves continually shocked when that neglect manifests itself; yet instead of facing up to the necessities of the human condition or the "human definition," we even more avidly embrace fantasies of the future as a problem-free state of existence.

Berry's notion of praxis then, turns on his understanding of what it means to live "within the human definition." This is more than a truism for Berry. However much "we long for the infinities of power and duration," living within the human definition means recognizing limits to our wisdom, foresight, and situation in nature (UA: 94). We can understand this by looking at energy use. Berry accepts that scientists can plausibly point out sufficient sources of energy to expand energy intensive agriculture. But even if we have "limitless" supplies of energy, "we can use them only within limits. We can bring the infinite to bear only within the finite bounds of our biological circumstance and our understanding" (UA: 84). Far from supplying needed moral boundaries, the visions of technologists and futurists only encourage us to deny that there are limits.

Pre-industrial societies did live within the human definition, according to Berry. Dependence on biological cycles restrained ambition. With cultivation came the need to store seed; ambition to sell or eat was disciplined by necessity. The user of animal traction was likewise restrained by the animals, by their wilfulness, slowness, need for food and rest. Such constraints developed more components of human character than just the will: intelligence, imagination, and good judgment (for continual decision making); nurturing, love, cooperation, and caring (because farming embodied the dependence of creatures upon one another); wonder, joy, and grief (at awesome things that kept happening); and humility and reverence (in recognition of one's smallness in the universe). It is these qualities (in balance with the will) that make meaningful life possible for Berry. These, not absolute perfection or infinite power, are what we are really looking for.

It is within such a condition that we can properly think of ourselves as free. Freedom for Berry is nothing other than living "within the terms and conditions of the real world" in a disciplined and responsible manner. He quotes Milton: "to be free is precisely the same thing as to be pious, wise, just and temperate, careful of one's own, abstinent from what is another's, and thence, in fine, magnanimous and brave" (*GGL:* 187). Hence freedom is not a matter of choosing uncritically from among the choices provided by the present culture—choices to buy and consume without thought of source or consequence, or to work without concern for quality of product and consequences of production. Central here is the possibility of a comprehensive self-mastery, which implies that there is effective recourse even when the conditions of the world would seem to block us from responsible living.

Berry's argument also has important consequences for the way we think about means and ends. In a cyclical world, in which we live our lives renewing cycles and recreating cultures, there are no ultimate, separate ends to aim at, beyond return. Ends blend into means, and appropriate means must embody the ends for which they are taken up. If our lives do not now exemplify responsible relations with land, family, and community, technical cleverness will never bring us to that end.

This approach to thinking about means and ends contrasts sharply both with the instrumental rationality of the Battelle report and Walters' justifications of the NFO holding actions. Both are founded in the idea that we are trying to get somewhere; to secure ourselves in the future we envision. With this linear view, a product of Judeo-Christian tradition, we have become so busy pursuing some future state of spiritual, material, and social perfection that we have taken no interest in living in the present and have forgotten to attend to the propriety of the means, Berry argues. It is as though we have arrived in our minds at a new age of fantasy or magic, we expect ends not only to justify means, but to rectify them as well. Once we have reached the desired end, we think, we will turn back to purify and consecrate the means. Once the war . . . for the sake of peace is won, then the generals will become saints, the burned children will proclaim in heaven that their suffering is well repaid, the poisoned forests and fields will turn green again. Once we have peace, we say, or abundance or justice or truth or comfort, everything will be all right (*D&H*: 130-1).

To Berry, taking such refuge in ends is evasive abstraction: "the discipline of ends is no discipline at all," for we can always invent ends either to justify past actions or to ignore present and future responsibilities. Yet "hope lives in the means, not the end," Berry insists; to sustain agriculture we do what is right, now.

The model figure of this agriculture is an old man planting a young tree that will live longer than a man, and that he himself may not live to see in its first bearing. And he is planting, moreover, a tree whose worth lies beyond any conceivable market prediction. He is planting it because the good sense of doing so has been clear to men of his place and kind for generations (*D&H*: 135).

A few pages later he reiterates: "A good farmer plants, not because of the abstractions of demand or his financial condition, but because it is planting time and the ground is ready—that is, he plants in response to his discipline and to his place" (*D&H*: 138). Our planting may lead to a good end, or a storm may wipe out our crop; we do not know which. But it is not even the possibility of good consequences that supplies the main reason for acting; the action is undertaken for its own sake, because it is the right thing to do.

If we need to think of ends, we should think of the proper end as health, not economic efficiency, the end usually used to justify projects. We justify turning a corn field into a shopping mall on the grounds that the mall represents a more economical use of the land. To Berry this is no justification at all—without knowing what the action is to be economical with regard to, we have no justification for it:

if we are going to think about the economy, we are obviously going to have to stand something beside it to measure it by. And it is clear that money is too abstract—and far too elastic—ever to make a reliable measuring rod. If we should, on the other hand, make health our standard, then we will have to ask some questions about an economy that, by the logical extension of its assumptions, causes catastrophic soil erosion and eliminates millions of farmers from farming by the device of economic ruin (Berry 1984: 29).

"Health," Berry notes, belongs to a family of terms including "heal, whole, wholesome, hale, hallow, holy" that together suggest a positive state of being that goes far beyond mere absence of disease. These suggest health is a comprehensive concept: one thing does not become healthy at the expense of another—"it is wrong to think that bodily health is compatible with spiritual confusion or cultural disorder, or with polluted air and water or impoverished soil" (*UA*: 103). A corollary of this view is that good solutions to practical problems do not involve tradeoffs; one does not trade spiritual or ecological health for bodily well-being, or present benefits for future problems. Instead, good solutions are what Berry calls "solutions of pattern." These "preserve the integrity of the pattern" (the pattern of the cycling of nutrients, for example) rather than "enlarging or complicating some part of a pattern at the expense or in neglect of the rest" (*GGL*: 144, 141). Good solutions are also small-scale, cheap, and avoid great risks.

We avoid tradeoff situations by keeping in mind the question "How much is enough?" The judgment of enough, like the determination to live within the human definition, is part of the discipline of community. The alternative to "enough," maximizing things, requires anyone's gain to be made up by someone else's loss and will continue to generate inequity, displacement, and poor quality work.

Finally, it is important to note once more that Berry has a quite different conception of historical process than Walters or the Battelle authors. For them there is a rationality to the course of events; the current distribution of people and resources (including the rise of cities) and the current range of institutions are seen to reflect inevitable responses to nature, technology, and society. The current social order must then be seen not as an accident incompatible with living within the human condition, but as a product of the human condition itself. In this view "you can't turn the clock back" because the social order and the human condition are continually co-evolving. We are products of our technology and adapt to it whether we choose to or not. As products of this historical dynamic, industrialization, urbanization, and capitalism must be accepted as legitimate and we must deal with their effects here and now; we cannot somehow step outside of the process of social evolution to reassert bygone standards and sensibilities. These historical conditions impose their own definition of "health"; to cling to earlier or other definitions only shows our incompetence to cope with the real world.

Berry, in contrast, sees conventional agriculture as trying to sustain an unworkable present by seducing us with fantasies of the future. He insists that all historical situations bring the possibility of living responsibly. Our institutions and the conditions of our lives are largely matters of our own doing, products of recent and ongoing choices, not the determined outcomes of historical forces. Responsible living may be harder for some than others, depending on how completely traditions of responsible cultures have been stamped out, but it is in principle accessible to all. Hence while there is undeniably change in history, there is no inevitable process which will take us to a certain spot in the future.

Berry—The Route to the Future

Because Berry locates the seat of responsibility in the lives of ordinary people, he finds radical change eminently practicable:

> The change I am talking about appeals to me precisely because it need not wait upon "other people." Anybody who wants to can begin it in himself and in his household as soon as he is ready—by becoming answerable to at least some of his own needs, by acquiring skills and tools, by learning what his real needs are, by refusing the merely glamorous and frivolous. When a person learns to *act* on his best hopes he enfranchises and validates them as no government or public policy ever will. And by his action the possibility that other people will do the same is made a likelihood (*D&H*: 123-4).

To those, like Walters and the Battelle authors, with differing estimations of the capacities and desires of individuals, this location of responsibility in the lives of individuals will be the weak point of Berry's argument. If individuals could act to produce the changes Berry envisions, surely they would have done so; we would not have the problems we have. Berry too is aware that the problem is not so simple; it will require great changes in outlook and values. In particular, we will need to learn to esteem the ordinary over the heroic and the particular over the abstract, and to celebrate good work and rootedness.

In an essay "The Gift of Good Land" Berry takes up the relations of the Judeo-Christian heritage to his understanding of responsible living. He notes that much of our literature, sacred and secular, extols the doings of heroes and that we so value the heroic that we have come to despise the ordinary. This has been a mistake; to Berry the ordinary is more worthy and difficult:

> The drama of ordinary or daily behavior also raises the issue of courage, but it raises at the same time the issue of skill; and, because ordinary behavior lasts so much longer than heroic action, it raises in a more complex and difficult way the issue of perseverance. It may, in some ways, be easier to be Samson than to be a good husband or wife [or farmer presumably] day after day for fifty years (*GGL*: 277).

A notion of industrial heroism has come to replace a notion of "good workmanship." In this "secularized version of the heroic tradition . . . the ambition [is] to be a 'pioneer' of science or technology, to make a 'breakthrough' that will 'save the world' from some 'crisis' (which now is usually the result of some previous 'breakthrough')" (*GGL:* 277-8).

Our preoccupation with the heroic is an indication that we have become seduced by abstractions. We solve great problems in our imaginations and find that so intoxicating that we establish institutions for constructing imaginary solutions. These are the institutions of specialization and they include many of the institutions of agricultural research. We have become good at coming up with abstract answers but have ignored the problems of their application. In agriculture "application (which the heroic approach ignores) is the crux, because no two farms or farmers are alike. . . . Abstractions never cross these boundaries without either ceasing to be abstractions or doing damage" (*GGL:* 280). The result is chaos: the culture that

> fails to provide highly articulate connections between the abstract and the particular, the organizational and the personal, knowledge and behavior, production and use, the ideal and the world . . . [leads to a] profound disorder in which men release into their community and dwelling place powerful forces the consequences of which are unknown. New knowledge, political ideas, technological innovations, are all injected into society merely on the ground that to the specialists who produce them they appear to be good in themselves (*D&H:* 127).

Part of the problem of living within the "human definition" is resisting the temptations of abstractions. But the alternative, living in the particular and ordinary, requires thinking differently about work. Work has been a target of heroic abstraction. We find ways to avoid work to pursue the fantasies we mistakenly call "the good things of life." To describe work we have come up with a concept of "drudgery": "a pseudo aristocratic notion . . . that one is too good for the fundamental and recurring tasks of domestic order and biological necessity" (*D&H:* 115-6; *UA:* 12). We cannot avoid this necessary work, Berry claims, and have only succeeded in making it uninteresting, unchallenging, and unenjoyable. But with thought, skill, commitment, and a recognition that what we were doing was necessary, the same jobs would be satisfying. We have examples of this kind of work in skill-enhancing farming practices, such as reaping with a scythe or using horse-drawn traction. Such practices limit speed and carelessness, encourage attentiveness, interest, and care. Working becomes something like raising children; one is concerned not with efficiency of production, but with quality of product and satisfaction in the experience. Certainly this attitude limits the amount one can do, but properly so, according to Berry: farmers should farm no more land than they can farm well. If we are content with "enough" rather than "ever more" we will be able to concern ourselves with quality. In doing so we will be exemplifying the ordinary, not heroically attempting to set records of yield, speed, and size.

Along with necessary work, commitment to place is the other primary means by which we live in the particular. We do not interact with the world, the ecosystem, or nature, but with a particular place. We may talk abstractly about our relations to nature or to humanity, but it is in the local community that those relationships are manifested. Berry makes much of Robert E. Lee's choice at the outbreak of the Civil War to sacrifice his antislavery principles and defend home and community: "his [decision] seems to me to have been an exemplary American choice, one that placed the precise Jeffersonian vision of a rooted devotion to community and homeland above the abstract 'feeling of loyalty and duty of an American citizen'" (*D&H*: 154). The example suggests how deep the commitment to community is to be. Principles are abstract and heroic, community is practical and particular. To stand firm for principle is undoubtedly noble, but if it cuts us off from community, our stand will have been futile. Our action will have come from a false sense of autonomy, a mistaken belief that we can ignore community and stand absolutely free (see also Taylor 1979). In practice, however, principles can only be applied within communities through the practical arts of community living: tolerance, conversation, negotiation and compromise, as well as confrontation. Communities may change more slowly than individuals, but the changes are likely to be more far-reaching.

Walters

Walters—The Future in Question

Walters hopes to bring us to an economic and ecological millennium. In it the institutions of scientific agricultural production will be both powerful and free from the influence of interest groups. Farming will become a great deal more scientifically rigorous than it currently is. There will still be family farms, but they will not be the mixed farms of the past celebrated by Berry and Logsdon, who think skilled, committed farmers can be relied upon to farm in an ecologically sound manner (Logsdon 1984: 3-4; Berry, *GGL*: 99-100). Modern farming (especially ecologically sound farming) has become too complicated to be left to farmers unless they are well-trained in eco-agriculture. This will not be the old "organic" agriculture of Rodale; it will use soil activators, which accelerate beneficial microbial processes, growth hormones, and catalytic growth regulators—many of the technologies *Agriculture 2000* envisions. These are not to be seen as the technical shortcuts with which we avoid questions of justice and responsibility, but natural secrets which will help us secure a more just social order. In Walters' view there is nothing wrong with technical solutions for moral problems, since the moral problems are in many cases ultimately avoidable by technical means; much of what is wrong with our society is due to a somatic illness, chronic malnutrition.

Implicitly eco-agriculture is concerned with the world as a whole: poisoning soil or people is not unique to any single country. Thus, the biological revolution must be international. Yet the social revolution must be national, since the main obstacles to eco-agriculture are bad economic and social policies that exploit farmers. In keeping with his populist heritage Walters strongly advocates policies to improve conditions for U.S. farmers: the cause of low farm prices is not overproduction but underutilization of agricultural products. We need tariffs to keep out cheap foreign produce and vigorous campaigns to expand markets for agricultural products. Walters denies the "inelastic demand" claim that has been used to justify the shrinking farm sector. It may be that there is only so much food we can eat, but we can find other uses for farm products (*HA*: 173).

These are details of the future we are headed for, but what does that future signify? Some of these details, after all—the information-intensive technological systems—resemble the predictions in *Agriculture 2000*. What distinguishes Walters' view is that his future is the final ideal state of humankind, not a stage in an endless series of adjustments. Walters view is also distinguished by its concept of the historical processes of struggle and discovery that bring us to that goal.

As we have hinted in earlier sections, Walters' view of history, both in the struggles of the NFO and in the necessity of eco-agriculture, is essentially millenarian, both in a weak and a strong sense. In asserting that there are right and wrong ways, based on natural laws, of conducting economic affairs and of farming and feeding ourselves, Walters presumes there is a possible condition of society in which we live in perfect accord with these laws. As we gain more and more true knowledge (not the stuff currently taught in universities) we get closer to realizing such a condition. Such a state of perfection is the millennium, and much of Walters' writing (and many contributions to *Acres*) exhibit a yearning for that millennium and the end to personal, social, and economic problems and injustices that it will bring. This is millenarianism in the weak sense, the notion of an achievable perfection of relations among humans and between humans and nature that can be sustained indefinitely. Walters shares this vision of an end to progress with many others who belong to the agro-ecological wing of the alternative agriculture movement. Those who see nature (including human nature) as a determined and lawful whole are implicitly assuming that there is an optimal way for humans to be and that once we discover it (which may not happen soon) we have the option of living in millennial perfection.

But Walters' writings are millenarian in a stronger sense as well. In common with Christian millenarianism he believes in the beneficence and wisdom of the creation, attributes what is wrong with the world to a combination of original sin and failure to follow the natural laws that make persons whole, sees cosmological significance in the struggles of an oppressed group, and foresees a final state of vindication, justice, and perfection. He sees need for and recognizes the existence of Messiahs, men like Staley and Albrecht, who recognize truth, lead for a while, yet are finally betrayed. His writing bristles with apocalyptic rhetoric and references to the eschatological tradition.

This millenarian outlook carries over into an understanding of the sacred directedness of history. We have pointed out already how the terms "Creator" and "Creation" are no mere metaphoric place-holders for Walters. These terms both symbolize and explain the orderliness of the world. Human history begins with a Creation, an initial laying out of laws and responsibilities, and an establishment of an ecological covenant which humans have with the Creator. Our "permission to life" is granted in return for proper management of the world and ourselves (*AP*: 395; *Acres* 7/82 39). As with Adam and Eve in the garden, we can choose to manage in the right way or the wrong way. With regard to our economic affairs, for example, "how man chooses to monetize new wealth from the soil determines his prosperity, political and economic stability. How he guards the fountain of that new wealth determines his historical rise and fall" (*CEA*: 6).

Almost always our choices are binary. We can do right and act in accord with the laws of creation, which is hard and requires knowledge, vision, and courage (qualities of the NFO leaders); or we can do wrong, giving in to our sloth, greed, and ignorance, and permitting class exploitation. Sometimes good has reigned, but sooner or later, human weakness corrupts society and ushers in an era of evil. History then, shows the perpetuation of struggle between good and evil, which turns out also to be a class struggle between those who produce wealth (agricultural produce) and those who would exploit them (*HA*: 123). "History reveals that old lessons have to be learned over again by each succeeding generation" (*AT*: 26).

There are indications in Walters' writings that an end is likely to come soon. Since late in the last century there has been an intolerable imbalance in the class struggle (*AT*: 24-6). Farmers are being displaced so rapidly that soon there will no longer be the critical mass needed to restore the balance (*AT*: 15). At the end of *Angry Testament*, Walters describes an America frozen into feudalism through its neglect of the social balance democracy requires:

> Once the corporate landlords have all the land, once they have gathered in the small merging forces and constructed great merging forces, then the era of peasant agriculture will have arrived, but not because sound economics demanded it. It will have been the masters of government speaking, and having spoken they will have brought ruin down on themselves (*AT*: 362).

That fate is not part of the inexorable working of things, but it is nevertheless coming fast because we are so seriously flouting the natural laws of economics. And all the while our economic policy has been leading us to the brink of irreversible imbalance (*AT*: 152, 5-6), bad food and environmental poisons have made us less capable of coping with that imbalance. Yet even while we are at the brink of calamity, there still remains the possibility—a realistic possibility—for salvation. The workings of the natural laws of economics give us reason for hope. As fossil fuel prices rise "even the most dedicated 'synthetic' farmer will have to do a double-take on this kind of technology. As he does so, his tillage program will have to change" (*AP*: 232). The attribution of human failings to incomplete food likewise brings with it hope for dramatic improvement. If we follow the truths that true science has demonstrated and eat well-grown foods, then we will have the vision and strength to act righteously.

Walters does not say which future will be ours. That is up to us. Here, as in other millenarian movements, the imperative to act is heightened by the enormity of inaction and the contrast between the future that could easily be ours and the grim way we're heading. To recognize apocalyptic and millenarian visions as symbols which motivate action, redress grievances, and unite people, in other words to understand them as myths constructed by desperate people, is no reason to take them less seriously. Instead, as Lawrence Goodwyn makes clear with respect to populism (1978), those millenarians who in conventional accounts may seem to represent an irrational resistance to reality may simply be asking for the justice the rest of society receives. Yet they are caught in a sort of rhetorical cuckoo's nest: the more forcibly they put their case, the more others see them as irrational extremists.

Walters—The Route to the Future

In a world of continual struggle between justice and injustice the only fit image of praxis is that of a just war. Recurring war metaphors in Walters' writing do many things. They give us a sense of common purpose, a notion of directedness, unity, and organization. We are to find meaning and worthiness not in accepting some "human definition" but in doing our bit for the common cause and under the unified command. They also simplify moral issues. There is an urgency to our action; the ends to be secured are too important to worry about means; it is quite right that we do what we must. These metaphors also suggest that struggle itself heightens vision and enlarges the vividness of our experience; our lives make sense, acquire purpose and importance through our participation in struggle. Each of these is important; we deal with them *seriatim*.

Order and Purpose. Walters' NFO books glorify not only the hero but the technology and organization the hero uses. The enemy we fight is a highly organized technical and economic system and it requires a similarly complex system to beat it. To win in the holding actions the NFO must coordinate the actions of farmers throughout the country. Its headquarters are "a nerve center." Inside is frenetic activity:

> Could NFO farmers in fact deliver on signal and on time? The ball went to the Marketing Area chiefs, and the chiefs took the problem to the members. Regional Supervisors wore their pencils down to inch-long stubs answering questions about FR and DFR personnel. Do they have county structure working? What progress as far as new members are concerned? Are they supervising county structure? How many teams are working each day? (*HA*: 110; *AT*: 217-8)

At the top is Staley, representing both the military and the prophetic bases of leadership—"commanding a personal loyalty usually reserved for a Bull Halsey . . . being looked upon by many as a prophet on a par with Elijah" (*HA*: 8).

The same metaphors appear in Walters' eco-agricultural works. Nature too, at all levels, is full of purposeful and organized action under a single command and in pursuit of a just end. The imagery is explicitly military and religious: "Much like an army patrol, the root tip pushes ahead, down into the soil, as if paving the way for a whole host of fine white hair troops looking for a meal" (*AP*: 26). Roots do not just grow, they act to obtain objectives. Nature is a painstaking commander: "Nature's programming tells the plant when to sleep, when to wake up, how to translocate nutrients, and how far to go in the food production business" (*AP*: 21). It is the existence of such orderliness and purpose that ensures the practicality of the eco-agriculture program. If we know that the parts of nature are programmed to work smoothly together toward a common end, we can discover what we should be trying to do by recognizing that end, and we can do it by decoding the complete the plan of the Creation and acting accordingly.

Ends, Means, and Morality. For Walters right and wrong are absolutes, with cosmic significance. Where the stakes are so high, there is little room for indecisiveness and none for moral relativism. Achieving the ends set up in the Creation is necessary for our survival, and this justifies whatever means we use to achieve them. During the meat and milk holding actions many people sympathized with farmers, Walters notes, and accepted their demands as reasonable, yet objected to the withholding of food from a hungry world. Walters' main response is that it is irresponsible to allow one's aversion to the means to interfere with the reaching of just ends. He writes of one stage of the action that "Even the old saw, 'they're going about it in the wrong way', vanished, largely because no one seemed to know what the 'right way' was" (*HA*: 222). For Walters, in contrast with Berry, the ends are not embodied in the means. We have somewhere to get to and we had better go the best way we can. It is simplistic, he argues, to make moral judgments on the basis of appearances without regard for the significance of the acts in question in the plan of the Creation (*HA*: 237). And even if we go further than appearances, we may still find that the rightness and wrongness of the means is a confusing, trivial, and divisive matter in comparison to the rightness and wrongness of ends.

Confrontation, Enlightenment, and Liberation. It is by participating in the cosmic struggle that individuals become free and find significance in their lives. Only crisis shocks us from our dullness. The holding actions, so incompatible with the ordinary sensibilities of farmers, forced them to understand the historical situation they were in. Their action liberated them from an Orwellian world in which platitudes and abstractions had replaced reality and disguised what was being done to them.

These ideas appear most clearly in Walters' discussions of the hog kills during the meat holding action of 1968. Farmers brought hogs (or cattle) to a central site, shot them, and buried them in pits. The visions and descriptions of the slaughter on television and in the newspapers horrified many and at first discomfited even the NFO members doing the killing. The slaughter was public, unceremonious, seemingly purposeless. Yet as the killing went on, its significance for the two groups diverged.

For the public it remained horrific spectacle, the mindless blood lust of an irrational minority. To Walters this outlook reflected the dullness of mind that sustained injustice: cows and hogs died equally mechanically all the time and were eaten without compunction; to be horrified simply indicated how far out of touch with reality a great many people were.

Farmers who killed their animals discovered, on the other hand, how enormously consequential their action was. Shooting one's own animals and dumping them into a pit so violated the image of a farmer as producer and husbandman that it could not help but shake the farmer into an understanding of how awful and yet necessary the commitment to justice had become. In this blood sacrifice then, so antithetical to the normal values of farming, farmers were asserting the intolerability of the existing market system and discovering their freedom from its domination.

Becoming free involves here a concurrent process of struggle and recognition. Once we understand the ecological and economic rules of the Creation we can see how pervasively "the powers that be" have perverted them. Without that understanding, we are not free, no matter how wide the range of options that appears to be available; and that understanding in fact narrows, rather than widens, our range of options. But to understand and not act is not freedom either, but simply an indication of inadequate understanding. So strong are the imperatives, that if we truly understand them, we are compelled to act; and as we act we realize, like the farmers, the full strength of those imperatives. Walters' freedom is a "Here I stand, I can do no other" sort of freedom. To be engaged in righteous struggle, no matter with what expectations of success, is proof that one is free.

In much of this Walters is in sympathy with Berry, for both see freedom and meaningfulness arising through the rejection of the constraints imposed by a disordered society and the acceptance of a responsibility grounded in nature. Both see liberation as deliberate action and both recognize that language can subvert action. They differ with respect to the accessibility of enlightenment and the circumstances under which it occurs. Berry finds the potential for responsible action in the commonplace activities of farming and family life. Walters, on the other hand, sees the acquisition of accurate insight as an uncommon gift.

The difference is important for it suggests how differently the two envision the process of social change. In Berry's view, real change takes place in homes, farms, and communities; in Walters', it must be imported from outside. Those few who know must instruct the many who are ignorant. Berry's conception is democratic; Walters' is technocratic. When Berry writes of reclaiming an eroded hillside, he is concerned with commitment and self-discovery; Walters offers rules on "How to Fill a Gully" in the pages of *Acres* (Bargyla Rateaver, "How to Fill a Gully," *Acres* 1/82: 14-16).

Battelle

Battelle—The Future in Question

As we noted earlier, *Agriculture 2000*'s conceptions of the future and how we are to get to it differ significantly from the conceptions of Berry and Walters. As social critics, Berry and Walters envision great change; their problem is how it can happen. The Battelle authors face a different problem. Their notion of the direction of the future comes from an extrapolation of trends and an analysis of the causes of those trends. Where we are now is where the past workings of economic forces have left us; where we go in future will likewise be where economic forces take us. As we saw in the last chapter, these forces are to some degree embodied in "agribusiness." Their net effect is also personified in a "market" that "*incorporates* all available market information and *adjusts* to changes in information and market conditions" (Battelle 1983a: 68, ital ours). Like a sentient and rational being, the market here acquires and evaluates information, and adjusts itself accordingly. The question of where we ought to go is therefore moot, and the consideration of deliberate and purposeful social change futile. At any point in the future we will be exactly where we ought to be and must be.

While this mechanism may seem abstract, the future it will provide is richly detailed in *Agriculture 2000*. It will include a great range of new mechanical and biological technologies that will increase productivity and solve many of the problems farmers face: new water-conserving irrigation systems, seeds treated with pesticides, plant (and animal) growth regulators and accelerators, new cultivars, salt-tolerant crops, hydroponic production systems, automatically guided tillage and harvesting machinery, alternative fuels, remote sensors and robot farming techniques, embryo transfer techniques for animal breeding, new animal confinement systems, and electronic means of making farming decisions and participating in markets.

Together these innovations will give us a future much like the present, only, so to speak, more so. Fewer farmers will farm more land with more, and more costly automatic devices and specialized inputs, yet food production will still keep up with population growth. Farming will no longer be so subject to natural hazards; it may even be possible to control completely the food production environment.

These changes constitute agricultural progress. But even if we agree that they are desirable, how can we be sure they will come about? How can abstractions like economic forces, even when based on generalizations from past behavior, produce such a definite future? If we're not sure how the actions we have taken as individuals got us where we are now, how can we know what to do to achieve this future? If humans are free, what is to stop them from realizing some other future?

One answer is simply to accept the report's predictions as assessments of experts and valid for that reason. But knowing that is not enough: we want to be sure that the predicted future is right and good. And if we are sure it is right and good, we need assurance that it will come, that we will not stray off into some other future. Equally, if we do not think it is right and good, we will want a reason why the alternatives we might desire are not possible. Hence, while the determinism of the Battelle approach and the detail of its predictions may make the question of *whither* moot, they make the question of *how* immensely important.

Battelle—The Route to the Future

The authors of *Agriculture 2000* draw frequently on the imagery of the free market and these images carry much of the responsibility for delivering us safely to the predicted future. In some passages the report does offer a variety of options for moving into the future; farmers and consumers are represented as free choosers (50, 57-8, 144-5). There are several allusions to the importance of market research in guiding innovation. Yet there is no claim that the detailed descriptions in the report are generated from current preferences as determined by market research; in many cases they seem to be drawn more from conceptions of technical possibility than from empirical knowledge of public preferences.

In representing the future as knowable and determined the report does not reject its concepts of market freedom; instead, it employs four chief strategies of reconciliation. The first of these is the invocation of economic law as a causal and selective agent, particularly through the metaphor "economic pressure." The second is the use of the term "trend" as a way of connecting what *has* happened with what *will* happen. The third is the use of a sentence format which may be represented: "As x happens, so y will come to happen." Such sentences take x for granted, and treat y, the innovation of interest, as a consequence or implication of x. The fourth means is the simplest: the future is guaranteed with the simple assertion that such and such *will* happen. We focus on the first two; only they shed much light on mechanisms of change. They turn out to be closely connected, for in some sense trends are seen as manifestations of economic pressure.

Agriculture 2000 assumes a constant pressure for change. Consider again the case of tractors. According to the authors, future tractors will carry computers to match engine speed to the amount of work the tractor is doing. Their engines will utilize waste heat and run on fuels, like ethanol, derived from plant products. They will be automatically guided by sensors, or a cable, or radio signals (115-121). But why will these changes occur? Farmers have traditionally been slow to adopt radical design changes in tractors. The answer is that they will come because of "pressure to continually increase the efficiency of the farming operation" (121). Likewise, for harvesting equipment, "the main factor affecting the rate of technology adoption will be the pressure to continually increase the efficiency of farming operations" (130).

In some cases the source of economic pressure is clear: "As the cost of labor increases [relative to other costs], the pressure to decrease the amount of labor used in vegetable and fruit production will increase" (130). Here the suggestion is that pressure for change can be deduced from laws of supply and demand with the degree of confidence one has in employing Boyle's gas laws. One may deny that the cost of labor is an independent variable that will continue to increase, but if one accepts that premise and sees the matter as within the province of economic laws, the conclusion follows.

But how can we be sure this pressure is translated directly into innovation? We need to know something about how the pressure is felt by the farmer or the manufacturer of farm equipment. We need some guarantee that innovation will relieve that pressure by reflecting the needs of users rather than, for example, the convenience of producers, the career aspirations of executives, the inspirations of inventors, or the need to keep the economy prospering.

In short, economic pressure does not alone explain whence comes the necessity to change or what ensures that the changes will relieve the pressure. For the most part, these questions are answered with assumptions: that there is permanent pressure for change and that there is some universal process operating that weighs the progressiveness of different options and allows optimal choices to be made. In one case *Agriculture 2000* does address the questions of what shapes technologies as they emerge: "ultimately, any new practice which conserves energy and labor, increases production, improves product quality, etc. must return the user a higher income than might be achieved from the conventional, perhaps less sophisticated, practice" (3). Rather than merely asserting that the more efficient process will be adopted, here the authors translate economic progress into its components, the factors that affect the farmer's net income. Yet if we look to this list to discover what factors shape the new technologies that are adopted, we may become confused. It is not clear what priority the items on the list have and it will be hard to avoid tradeoffs among them—especially between labor and energy, or quantity of production and quality of product. Thus, while effort to improve in any of these areas may be seen as desirable, the list of variables does not itself determine what technology will increase income. And the term "ultimately" at the beginning of the passage raises even more clearly the problem of what is to shape the technological future. To claim that the ultimate achievement of an increased income determines the practices used to achieve it is to make the future the cause of the past. But we must choose now; we have not the luxury of waiting to see which choices are ultimately wise. The statement may be true but it offers no principle for innovation. Let us go on to examine "trends," where we may find a more illuminating way to understand how economic pressure can give us a definite future.

Almost every attempt to predict the future utilizes trends, whether or not they are ever explicitly articulated. But selecting, measuring, extrapolating, and assessing the meaning of trends is a tricky matter. We must decide what phenomena are worth measuring, how to measure them, how to separate signal from noise. Our data may be too few to make clear whether change is linear or exponential. Even if we know this we will need to decide whether the phenomenon is peaking at the top of a bell curve or leveling off temporarily in preparation for a sharp rise. We may hope to answer this by inquiring into the complex factors controlling the trend. Finally, to predict the future we must assume that we have uncovered the relevant trends and what generates them, and recognized their effects on one another.

Here we leave aside most of these problems, recognizing that any attempt to know the future will need to come to grips with them. As the popular digest of a technical report, *Agriculture 2000* leaves out much of the data used in constructing the trends it identifies. We focus instead on the way "trends" are used rhetorically in the text. We raise such questions as "What kind of thing are these 'trends'?" and "What potency does 'trend' have as the subject of a sentence; what, in other words can 'trends' do?" These questions are important for three reasons: first, because a trend links the choices of individuals with the cumulative effect of those choices. Second, the identification of trends is a way of reducing possibility from "anything might happen" to "this is likely to happen." Finally, in identifying trends, *Agriculture 2000* makes clear what individuals are to do in contributing to those trends; what, that is, our praxis should be.

One of the things the language of trends does (and a key part of its potency) is to overwhelm the significance of individual events. Trends, after all, are expressions of averages; they supersede the individual events that make them up. The authors write of the trends in farm income, for example:

> Trends from 1950 to 1980 in gross and net income from farming are shown in Table 6. The cost/price squeeze is currently being encountered by American farmers. This is exemplified by the declining trend in net farm income as a percentage of gross farm income which equaled 41.2 percent in 1950, 29.6 percent in 1960, and 24.2 percent in 1970, and fell to 13.2 percent in 1980 (12).

For Walters or Berry, these statistics would represent innumerable human tragedies whose sum (not whose average) would manifest a great injustice. Here by contrast, there are not only no individual events, but no experience and no agency, hence no opportunity to place blame. The key verbs are passive—"trends . . . are shown"; "squeeze . . . is being encountered."

Depending on where one locates hope, this image can either be threatening or comforting. It is threatening for Walters: it alienates people from experience, hides conflict in statistics. Yet the portrayal can also be comforting. Because it refuses to grant significance to the accidents of individual experience, it reassures us that we are safely passing through difficulty—we need not worry, because our world will adapt. Much of this rhetorical work is done by the sentence dealing with the "cost/price squeeze." If the "squeeze" really "squeezed" we might better say that "the combination of high costs and low prices is squeezing the profit out of American farming" or even "squeezing out the lifeblood" or "strangling" the American farmer. "Squeeze" could bring with it other metaphors of violent physical force. But the verb, "is currently being encountered," draws in a quite different metaphorical tradition. We "encounter" phenomena as we journey from one situation to another, through time if not also through space. To think of the "cost/price squeeze" as a transient event on a journey into the future implies that American farmers will pass beyond it. No matter how grim the immediate prospect—however hard the "squeeze" may be—it is tolerable if we know there is a future beyond it. This is exactly how *Agriculture 2000* views its present:

> There is little doubt that U.S. agriculture is going through a restructuring period that will affect its future over the next two or three decades. The problems of the future will not be any less complex than those faced today. However, the farmers and agribusinesses that emerge from the current

recession should be stronger than ever, and better equipped to face the challenges of the future (19).

The passage does not promise that all will get through, though it does invite each to be one of the survivors. It does not promise ease, but because survivors will be "stronger" and "better equipped," the "challenges" of the future will be less troublesome than those of the past.

Thus, in confusing or difficult times, this impersonal language of trends may help to neutralize the crisis of individual experience by submerging it in the total experience of a group. It can provide us with reason for hope by giving us a line on a graph—a road on a map—to follow into the future.

Trends reduce possibility because the term comes to have normative significance. In part this comes from designating as trends many types of time-series phenomena, with quite different causal characteristics. In some cases trends are simply empirical, matters such as consumer "tastes and preferences," which may not admit of further explanation. Other trends are clearly determined by independent social or demographic events. For example, a downward trend in the number of new college graduates having farm backgrounds can be understood as a consequence of the declining farm population. The idiom of trends (though not always the term itself) even arises in cases where phenomena may be presumed to be consequences of physical laws, such as the claim that increased demand for agricultural products will produce an upward trend in soil erosion through the increasing cultivation of marginal land (34).

Identifying a trend enhances the status of the phenomena that make it up; they become something more than isolated accidents (see Figure 11.2). And as the examples above suggest, the trends themselves are not just generalizations, but represent the normal, perhaps even the necessary. Anyone wanting to promote an alternative must bear a burden of explaining why an existing trend should not be. *Agriculture 2000* notes, for example, of the trend of livestock production to keep pace with U.S. population growth, that "nothing is foreseen in animal agriculture technology to indicate that this trend will be reversed in the foreseeable future." Here what needs explaining is what might change the trend; no explanation is needed for its continuance.

Finally, in *Agriculture 2000*, trends not only show us where we are going, they get us there. And if they are the tracks that keep us from going off into the wilderness of possibility, they are also the driving force that moves us along the track. They do this because they acquire both a moral and a physical momentum.

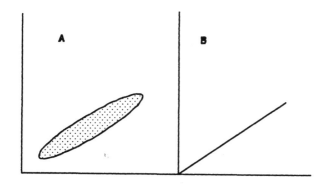

Figure 11.2 How Empirical Data (A) are Reified as a Trend (B)

Trends possess a moral momentum because they become standards against which individual behaviors are judged. To reify a set of behaviors as a trend is to make the mean of those behaviors normative. Some of the individual behaviors that make up the trend may be so far from the mean as to be judged *deviant*. But we give terms like "norm" and "deviant" double meanings: they describe both the qualities of individuals in relation to a group and the qualities it is or isn't desirable for individuals to have. When we reify a set of phenomena as a "trend" we give it a capacity to compete for moral priority with the individual behaviors that make it up. For example, *Agriculture 2000* notes of changes in marketing that "many of these changes will be extensions and increased acceptance of current trends" (63). Here trends are no longer just generalizations: people are put in the position of being expected to accept and carry on trends. Those who have not accepted are presented as at fault, behind the times, resisting reality (see Figure 11.3).

Trends also possess a quality analogous to physical momentum in *Agriculture 2000*. A trend comes not just to have a kinetic signification, but an inherent dynamism; it represents a social vector. The authors write for example that "although total confinement systems for hogs are not widely used, *the trend is rapidly advancing* as new pork production systems become larger" (157, ital ours). Here the trend itself is pictured

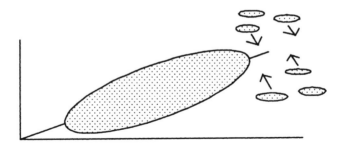

Figure 11.3 How Trends Become Normative:
  Once a trend is held to exist there can be normative pressure (arrows) to
conform to it.

as moving to meet an objective. It has velocity and direction, it is even
accelerating. It takes us to a particular future and away from other
futures (see Figure 11.4).

References to "future trends" are one of the clearest means by which
the report guarantees a particular future. The authors write that "future
trends in animal agriculture point toward larger and more sophisticated
units, with increased emphasis on confinement techniques" (135) and "the
future trend in beef production is toward a 'genetic package' of beef
exhibiting uniform sex, weight, age, genetic background, and feed
efficiency characteristics" (143). In one sense, "future trend" is a
contradiction; trends are time-series data of the past that can be extrapo-

Figure 11.4 Two Ways of Moving into the Future

Pushed by Pressure

Riding Trends

lated into the future but not drawn from it. Yet the concept does do rhetorical work. Consider, for example, the claim: "because total meat consumption per capita has remained stable in the United States over the past decade, various components of the animal agriculture sector (beef, pork, poultry, and fish) will be competing with each other for the consumer's dollar" (57). Here a ten year trend is the basis for denying the possibility of a significant increase in meat consumption. The trend, not consumer choice, resource constraints, policy changes, or anything else, determines the future.

Another example of trends directing possibility is in the discussion of small farms. The authors write: "Certainly the trend toward increasing sophistication in livestock production and the associated higher capital costs, associated with more confinement raising of livestock, makes return to more diversified agriculture somewhat difficult (61)." In this sentence the difficulty of making a change resides in the trend itself, not in any of the particular institutional and technical factors that might make up the trend—everything from the cheap food policies of governments to the lack of appropriate machinery. To give to a trend this kind of power suggests that the term here does carry the moral legitimacy and momentum described above. If the trend reflects changes that are somehow understood to be right or necessary (perhaps because they are determined by economic pressure) then it would be wrong (and probably futile) to oppose it. If the trend is seen to possess a momentum of its own, beyond the sum of the individual factors that make it up, then an even larger effort will be needed to change it. One will have to stop the trend as well as its various components.

Mainly what the language of "trends" does, like the language of progress, is to provide assurance that all is going well. Things are changing in a particular and appropriate direction, one that reflects economic rationality and democratic process. We have left diversified farming behind not owing to some chance association of accidental factors, but of necessity, and for reasons that are legitimate. Trends assure us that the choices we have made and will make are the right ones.

Associated with the view that we choose wisely is the view that, however complicated things may seem, our future is not really in jeopardy. Because trends, like decisions in a free market, embody both free choice and goodness, there is no risk. No fundamental policy choices confront us in which the choice made might send society in a very different direction from that which it needs to take. At worst, we may temporarily fail to keep up with reality; if so economic force will quickly boot us in the backside and we'll catch up. This includes problems that figure prominently in the works of the other authors: the potential for soil erosion, shortages of irrigation water, the welfare of animals, farm laborers, and family farmers. On these issues the choices we face are either illusory or insignificant: whatever we decide, the future will come out pretty much the same. We may feel our sense of cosmic responsibility lightened even if we are not quite persuaded that our concerns are idiosyncratic, ill-advised, and transitory.

Without the possibility of making meaningful changes there would seem little opportunity for raising the issues of praxis. Yet the report is in fact advocating a praxis: it is asking for trust. In return its makes a promise:

> Given time, and the *combined effects* of new developments to increase farm productivity, the problem of resource availability should not have a significant impact on overall U.S. farm production by the year 2000. This is not to say that there won't be problems and economic dislocations at some local levels where water shortages and soil erosion may be especially acute (50).

The promise here is of abundance. Production can and will meet demand. There will be problems for some people along the way. But because the problems don't threaten the outcome, they are no more than bumps in the road. We can sit back and enjoy the ride.

Each of these authors has hoped to communicate a vision of agriculture and social change. And each has succeeded, finding a large audience in those who share sensibilities, outrages and hopes, and have similar notions of how things work. For those who share their outlooks, much of what these authors have written needs no justification or explanation: Berry doesn't bother to explain why he is so insistent on his particular notion of a meaningful and honest life, or why air-conditioned tractors with sophisticated stereo systems are so much an anathema. Nor does Walters have to justify to his readers the malevolence of the lineup of antifarm institutions. And the audience of the Battelle report will need no convincing of the virtues or necessity of the kind of mixed market system that prevails in American agriculture; it represents, after all, nothing less than a response to social and natural reality.

That we, and we hope most readers, can feel something of the impact of all these persuasions, that some of their obviousness comes across, is surely an indication of the common culture of America that we bear. But what happens when people like these authors take their very different ideologies away from the context of conversation with the converted and apply them in broader public forums? One result is confusion. The available reasons that any of our ideologies could mount in principle are so complex that even if we were fully in touch with them, we would have an exhausting job presenting them in a policy context (to say nothing of defending them).

It is perhaps for this reason that political decision making so seldom probes the depths of disagreement. To record opinions and desires is enough; the sensibility from which they come is rarely inspected closely. Yet political discourse can appeal to reason: we can and sometimes do debate, inquire, and deliberate at length to get better policies, and we try to take ourselves seriously in doing so.

The next two chapters take up issues of what happens to these ideologies in a broader public where none simply dominates and where the business at hand is not preaching to the converted but demands instead that we persuade others who are not like-minded. We look first to the legislative process, particularly the hearings stage, where in theory public input is most directly brought to bear on the content of legislation. There we find a hurried process that deals with grave issues by cliché and euphemism. Some participants have learned vocabularies that neutralize ideology; others, refusing to adopt such language, find themselves simply unheard. Little communication occurs, not necessarily because the hearing is a disingenuous event, but because the communication techniques at hand are so vastly inadequate to what many hope to communicate. We turn then to agendas for agricultural research. There we find attempts on all sides to preempt the long and risky process of policy choice by the shortcut of "discovered" truth. The research agendas are less a matter of questions to be answered than tasks to be done. The ambiguous character of agricultural research as both science and development permits this; if we can discover or invent the world we desire (remembering always to observe that it is natural and progressive), we will never have to go through the messy process of convincing others that it is a good idea; if we can take control of the research agenda it will soon be a *fait accompli*.

PART THREE

Getting Closer to the Action

12

Ideologies in the Legislature:
A Guide to Hearing Hearings

One basis for the claim that U.S. agricultural policy is democratically responsive is the system of congressional hearings. Each year a great range of views, from those of trade associations to those of public interest groups, becomes part of the public record of congress's confrontations with an equally large range of agriculture related issues.

Political scientists write about hearings in two contrasting ways. On the one hand, hearings are to be a means of communicating knowledge to members of congress. Hadwiger and Talbot write that "public hearings are supposedly to educate committee members as to the substance of legislation and how groups feel about it" (1965: 119). In the Plant Variety Protection hearings we consider below, Senator Donald Stewart opens by claiming an "open mind." He tells witnesses that he looks forward to "expanding my knowledge" and wants their help to become the Senate's expert on the issue (U.S. Congress, Senate 1980 [S]: 2, 11, 33). The term "hearing" itself would seem to imply an openness, an admission that legislators need to be educated, and perhaps a notion of legislation as the reconciliation of conflicting views into informed and maximally acceptable policy.

A contrasting view sees hearings as exercises of raw power. Hearings, writes Bertram Gross, are "staged"; they are part of the spectacle of politics rather than its substance. Minds will already have been made up; members will use hearings to bring out "facts and opinions favorable to their point of view" (Gross 1953; Hadwiger and Talbot 1965: 119). Hence little hearing may go on; indeed, it may be the witness who is made to hear the legislators (who are doing the "grilling"). Getting one's views on the official record may be the only outcome of testimony.

Other perspectives on the process are possible; "hearings" covers a great variety of inquiries. And in the best of circumstances the effects of one's testimony may be untraceable; it may be lost among other testimony, and subject to compromise, apathy, and the pressure of business as well as falling victim to outright opposition.

Here we suggest another way of understanding what occurs in hearings on agricultural matters: members of congress, with ideologies of their own, and steeped in congressional tradition, briefly encounter witnesses representing a wide range of ideologies. Differences in outlook may be so wide that communication is subverted, even where there is genuine effort to communicate. Members of congress will have difficulty focusing questions to get at the heart of the networks of understanding that underlie witnesses' concerns; witnesses, limited to brief statements and responses, will be unable to project what to them is self-evident in their own assessments. This view of hearings as clashes of contrasting understandings does not exclude the first two views. As a third way of seeing hearings, it can help to reconcile them by indicating how the same event might appear both as information gathering endeavor and exercise of raw power.

"A Minor Bill Freighted with Controversy": The Plant Variety Protection Act Amendments, 1979, 1980

The case we examine here, the House and Senate hearings on the Plant Variety Protection Act Amendments held in 1979 and 1980, is particularly suited to our approach. In 1970 the Plant Variety Protection Act became law. It provided a means for granting patent-like certificates providing seventeen years exclusive sales rights to breeders who developed new varieties of nonhybrid plants. Soup industry opposition had kept peppers, okra, tomatoes, celery, carrots, and cucumbers off the original list, but with industry cooperation, the USDA moved in 1979 to amend the act to include these vegetables (and to extend the certification period to eighteen years in accord with international conventions). It met fierce opposition from a number of public interest groups: The National Sharecroppers' Fund, The Consumers Federation of America, The National Association for Gardening, The Environmental Policy Center, The Environmental Defense Fund, The National Center for Appropriate Technology, The People's Business Commission, The National Farmers Union, The Center for Rural Affairs, and The National Coalition for Development Action. These objected not so much to inclusion of the six vegetables as to the idea of patenting plants. Although unrepresented, the John Birch Society was also a source of grassroots opposition, an indication that objections came from the far right as well as the environmentalist left. Supporters of the bill included the USDA, the seed industry as represented by trade groups, The American Patent Law Association, several academic agronomists and crop scientists, and, perhaps surprisingly, some elements of the organic gardening community (DeCrosta 1980). The bill did pass congress at the close of the Carter administration.

The hearings took place over the span of nearly a year. House hearings were held in July 1979, with ten witnesses, seven supporting and three opposing. In April 1980, Representative E. (Kiko) de la Garza resumed the hearings, noting widespread "misinformation" distributed by "people who were completely uninformed" with regard to "possibilities or future possibilities, up to and including bringing about the end of the world if this bill were to pass" (U.S. Congress, House 1980 *[H]*: 80; Doyle 1985: 68; Wehr 1980a,b). Of sixteen witnesses in the second series of hearings, eight were in favor and eight opposed. A great many letters from small seed companies supporting the bill were also presented. The unexpected controversy the bill had generated and the widespread "misinformation" (mostly from the John Birch Society) led de la Garza to conduct these hearings in the context of due process rather than education of congress. The hearings were "to satisfy everyone who has voiced some concern, . . . not wanting at all for anyone to feel that they [sic] were not heard" (*H*: 80). But while the concerned parties were to receive the gift of access to the legislative process, it was as much to have their errors corrected as to have their views heard.

The Senate hearings took place two months later in June 1980. Of ten witnesses, five were in favor and five opposed. All had testified at the house hearings and they made much the same arguments in the Senate hearings. In contrast with the house record, the appended correspondence was overwhelmingly opposed to the bill.

The hearings were neither forums for acquiring knowledge (except in a loose sense) nor orchestrated displays of power. While it may be argued that both committee chairs (Rep. de la Garza and Sen. Stewart) were favorably disposed to the bill (de la Garza was a sponsor), the legislation was too minor to warrant much effort to manipulate the hearings. Indeed, the hearings were less display of power than exhibition of perplexity: both the two chairs (who asked most of the questions and were often the only legislators present) and the congressional press could not understand why a minor amendment generated so much fuss (Wehr 1980a). Witnesses, pro and con, were rarely able to ground their positions in fact. To be sure they gave reasons—inferences from the laws of nature or economics, assertions of human rights and human nature, assessments of the state of the world and its prospects for the future—but these were little more than fragments of ideologies that were never exposed fully enough to be carefully examined in the hearings.

The perplexity (and impatience) of the questioners was most visible with respect to the scope of the hearings. Critics of the bill, some of whom were disturbed that the 1970 act had passed with little debate or oversight, insisted the issue was much larger than six more vegetables and one more year. They wanted to talk about the goals of seed-breeding programs, monopoly in the seed industry and its impact on farmers at home and abroad, and the maintenance of a diverse gene pool. De la Garza treated the raising of these issues as a matter of misunderstanding. Those (e.g., the Birch Society) who saw the bill as leading to takeover of "the United States . . . by controlling its . . . food and fiber" were simply in error (*H*: 82). A witness concerned with threats to global ecological stability was told that he had "missed the . . . boat" in bringing this up (133). (Stewart was more willing to expand his hearings, asking to be educated in these matters but expressing frustration that witnesses could not supply the facts he wanted. He threatened to hold up the bill until the germplasm protection issue was satisfactorily addressed [*S*: 2; Doyle 1985, 46-9]). The opponents succeeded in broadening the agenda, if not in making their point: de la Garza questioned them (and sometimes lectured them) about their concerns; proponents tried to explain why their fears were groundless.

The Issues

There were several objections to the PVPA amendments:

1. They would raise seed prices—ostensibly this was grounds for the opposition of the National Sharecroppers' Fund, though in fact its lobbyist, Cary Fowler, had longstanding concern about another issue, the loss of germplasm. Defenders argued that price increases were more than compensated for by increased productivity.

2. They would lead to overconcentration in the seed industry. Critics argued that certification would make (and had made) small seed companies prime targets for takeover, leading to monopolies in commercial seed for certain vegetables. Especially worrisome was the prospect of takeover by agricultural chemical companies, which might favor development of more chemically dependent varieties. Defenders claimed certification would give smaller companies a better means to compete: control of a unique variety would counter the sophisticated marketing techniques of larger firms. They also argued that takeovers of small seed companies since 1970 reflected general business trends not the influence of the act.

3. They would lead to genetically uniform crops, which would be especially vulnerable to pestilence. Opponents argued that monopolization of plant breeding would increase pressure on farmers to grow the same new varieties and that what was needed was a vigorous effort to preserve diversity, not protect the investments of private breeders. Defenders responded that certification and germplasm preservation were separate issues. USDA representatives claimed that because certification brought with it the need to identify and store germplasm, it actually made protecting traditional varieties more feasible.

4. They would lead to regulation of the seed market. Opponents feared the 18 year provision was the first step to the U.S. joining UPOV, the International Union for the Protection of New Varieties of Plants and hence to the European practice of allowing sale only of registered varieties. Some believed (mistakenly) that the bill itself would prohibit farmers and gardeners from growing uncertified varieties. Defenders pointed out that the bill contained no restrictions on uncertified varieties and that wide variation of growing conditions in the U.S. made UPOV-style regulation of seed markets impracticable.

5. They would inhibit communication among plant breeders. The prospect of proprietary control over new seed varieties would make geneticists less willing to share sensitive information with each other, critics argued. Certification might therefore slow progress in plant breeding. Supporters noted that germplasm of certified varieties would be open to use by other breeders. They argued also that certification would lead to more research, which would compensate for any hindered flow of information.

There were other, broader issues—the right to claim ownership of a life form, the seriousness of plant extinction or loss of habitat, the desirability and viability of high-tech monoculture. The act's ironic title did not escape attention: the "Plant Variety Protection Act" was, as a supporter put it, to "protect *developers* of novel plant varieties" (*H:* 76, ital ours). Far from being protected, plant varieties might be made extinct by the bill, critics argued.

For the most part, the desired ends were not at issue; both sides were concerned about genetic vulnerability, the viability of small farms and small seed companies, the need for many varieties of commercial seed, and the farmer's freedom to choose what seed to grow.

Precisely what the various parties hoped the hearings would do was not always clear. The members of congress appeared to have had no particular case to make; the bill was too minor to have significant political implications. Their investment was in correcting misunderstandings, assessing the range of criticisms, and discovering the political, economic, and social implications of the bill. The witnesses (both pro and con) hoped to establish their versions of reality and, especially as the hearings wore on and the range of disagreement widened and deepened, to dramatize the risks of failing to recognize their realities. Opponents had the harder job: the bill was technical, made only minor changes in the law, and was supported by the USDA and by the small seed companies that might have been expected to oppose it. They may have hoped that the inertia of legislative process would magnify their objections, or they may have felt a moral obligation to testify. They may have believed their concerns would be met sooner or later in other bills, or that their truths were powerful enough to induce an immediate change in plant patenting laws.

Fighting over What Is Real

There were in fact three distinct parties in the hearings: opponents, proponents, and legislators. Each claimed to represent reality. For proponents reality made it imperative to pass the bill, for opponents reality made it imperative not to pass it. Yet legislators too called on witnesses to "deal with the real world," as Stewart tactlessly put it (S: 57). Along with listening and learning they periodically made ex cathedra statements about human nature, economic law, or justice. Their reality was much more than Capitol Hill *Realpolitik* (though that was present), and it wasn't simply a reflection of their views on the amendments either—de la Garza and Stewart, sponsor and skeptic respectively, commanded their hearings in similar ways.

These congressmen confronted witnesses ideologically, in the sense we developed earlier: they "knew" how the world worked, what was politically possible, what it meant to be a representative, and what they needed to hear in the hearings. But how was their reality brought to bear?

In one sense the legislators already controlled reality. They controlled legislative possibility: at the end of the hearings they, not the witnesses, got to say what the event meant. At the same time, however, witnesses did have influence; their conversations with members of congress revealed a degree of tolerance and a willingness to be corrected and convinced on both sides. And legislators took pains to demonstrate that their claim to represent reality was warranted: that they saw further, thought more sharply, represented more interests and concerns, and acted more appropriately than witnesses.

The legislators' claim to represent reality was in fact three claims. First was a claim of cognitive authority, that they knew more about the world than witnesses who had causes to plead or specialized insights to offer. Second was a claim to represent a broader constituency, a claim that their reality reached out from Washington to take in more interests—their own constituents, other Americans, virtually the future of the world. Third was a claim of open-mindedness, a claim that they were more open to reality because they were the learners. It may seem that these claims are not easily reconciled, yet both hearing chairs invoked each of them regularly. Together, they constituted an assertion of judicious wisdom: that legislation acceptable to such an individual would be consistent with general truths, reflect concern for (and even protection of) the rights and welfare of all, and be grounded on the facts of the matter.

Let us take these claims in turn. First the cognitive authority claim. An opponent of the bill, Ginger Nelson of the National Center for Appropriate Technology, complained that large companies no longer paid attention to consumer interests. This de la Garza denied: "but that is what they live from. Without a consumer, there is just no corporation out there. . . . Everywhere I go, they cater to the consumer. . . . you cannot make the profit unless you have a satisfied customer" (H: 95). During the Senate hearings Barbara Schlei, a USDA official and supporter of the bill, argued that antitrust laws would prevent undue concentration in the seed industry. Stewart lectured her: "Hogwash. Those things [antitrust laws] have not been used since they were put on the books, not effectively, not successfully; . . . that is just hogwash. I am sorry, but if the antitrust laws are relied on to weed out concentration in the economy of the agricultural sector of our economy, or anything else, you could not depend on those if your life depended on it" (S: 17).

Both responses have a "C'm'on now, what kind of fool do you take me for" quality. Both are put forth as final. Each presumes a superior familiarity with reality, but neither invites an empirical test or a thorough analysis of a complex political-economic environment: the guarantee of their truth is the authority of the speaker. De la Garza simply pulls out a commonplace economic principle and insists that Nelson accept it. Stewart presents himself as being in a privileged position in which he can see a grim political reality. Yet he offers no evidence that antitrust laws never work or cannot work. The vehemence of his claim—"hogwash"— leaves Schlei no viable response. He dismisses her suggestion that antitrust laws may need to be strengthened, and she finds relief only by admitting that the amendments will not, after all, be anticompetitive (*S:* 18).

The second claim, of broader representation, was also a claim of paternalism. It involved imagining a constituent of a particular social status, occupation, and educational background, and reflecting on the effect the legislation would have on that person. It was not so much a response to grass roots input as a demonstration that all were being taken into account. Even when reference was made to a particular letter a legislator had received, the writer of the letter became a caricature to be manipulated. De la Garza asked a proponent to acknowledge that "that little farmwife out there that someone has gotten to write to us saying that if H.R. 999 passes, she is not going to be able to have her little garden in the back yard—she does not need to worry too terribly much about this, does she"? (*H:* 174) Treating the problem as belonging to an individual symbolized concern for all people as individuals, even though de la Garza used the letter for a purpose opposite to what its sender intended. Thus in the guise of concern for each citizen, de la Garza privileged his view as a member of congress over that of an unnamed woman in the provinces. The issue here is not who was right—de la Garza was likely correct that this person misunderstood the bill—but the way in which claiming to speak for those not present involved elevating one's own reality over the reality of all others. It is, of course, a central claim of representative democracy—we elect the people we think embrace reality as we see it and we trust them to do just that (*H:* 80).

The frequent use of the term "little" belongs to the attempt to broaden the representativeness of the hearings. De la Garza referred to "the little peasant out in the desert" (*H:* 184); Stewart to "the little old farmer" (*S:* 12). Both hearings were run with a controlled informality, the chairmen sometimes displaying a self-deprecating banality. De la Garza joked about the hotness of his district's chili peppers, for example. Such was the context of the "little farmwife." De la Garza followed up the remark by putting himself in the same situation: "I am going to be able to have my little chili patch, and she can have her garden?" he asked a witness to confirm (*H:* 174). In belittling themselves the chairmen were building bridges between themselves and their constituents, isolating the witnesses with their arcane knowledge and special interests, from all the rest of us, who are just plain folk down deep. The bridges gave credibility to the claim that the congressmen spoke for the ordinary person and thus could take an elevated position of authority. Hence "little" was not a derogatory term but an assertion of democratically based power.

The third element is open-mindedness. Questioners presented themselves as empiricists; the hearings were for them to learn the facts. Stewart, expressing at the outset his ignorance of the issues, pledged to become the Senate's expert on germplasm preservation. Again and again the chairmen represented themselves as learners: "You tell me." "Explain it." "I don't understand" (*S*: 32, 34; *H*: 14). To be seen as open to truth was clearly necessary if one was to claim to represent reality. Yet expressing ignorance did not mean ceding one's power to declare reality to the witnesses who gave the information. Statements of open-mindedness came with demands that witnesses convince skeptical congressmen of the worth of their testimony. Stewart was a particularly aggressive questioner, often calling for more detail than witnesses could supply (*S*: 16, 26), or for proof that what they claimed as causes were not merely correlations. He went out of his way to discover the coherence of witnesses' positions, sometimes explaining that while he was sympathetic, it was necessary for him to be "devil's advocate" to insure "a good record" (*S*: 43). And failure to follow up with such interrogation did not mean statements had been accepted; it often reflected a decision that they were not worthy of any more time.

Substantive Differences

Let us turn now from form to content. We focus on the conflicting views of proponents and opponents on two themes. The first is their presentation of their views in terms of rights, economics, and politics: their depictions of the existing economic situation, of where economic forces would take us were the bill to pass or fail, of what futures were possible, and of what actions were needed to get us where we had to go (see Figure 12.1). The second, narrower though more complex, is their response to the vulnerability to pestilence that came with increased genetic uniformity among food crops.

Generally witnesses agreed that developing new plant varieties was a good thing. They disagreed about the social circumstances in which plant breeding should take place and what forces would guide it to desired ends. Both sides founded their cases in conceptions of market economics and human rights. With many Americans they shared the belief that a truly free market place was justified as an expression of the human right to free choice, and could be relied on to improve general welfare.

Proponents emphasized property rights and systems of market incentives. Plant breeding was but another form of intellectual creation, deserving the same protection given to inventors, writers, or artists, they argued. William Wampler of Virginia, the most conservative member of the House committee, declared it a "fundamental right" of an inventor to own an invention (*H*: 192). Charles Lewis, former Executive Secretary of the National Plant Genetics Resources Board, argued that the 1970 act had been passed "as a matter of justice and equity" (*S:* 43). To these people the act corrected an injustice that had grown increasingly serious as plant breeding had become more time-consuming and difficult.

The rights argument blended into a utilitarian argument that granting rights to breeders would ensure sufficiently vigorous breeding efforts. Wampler saw patenting as the best way to meet anticipated demand for highly productive varieties. Billy Caldwell, a North Carolina State University soybean expert and President of the Crop Science Society of America, reported that plant breeding no longer attracted the best minds; it was too tedious. The incentive of market monopoly would make it worthwhile. Making plant breeding a private enterprise would allow public sector researchers to concentrate on fundamental work that did not have a direct payoff (*S:* 40).

Opponents repudiated the intellectual property analogy (*H:* 186), but they did accept that widespread development of new varieties could enhance real pluralism in agriculture and thereby mitigate the threat of catastrophe due to genetic vulnerability (*H:* 93). But this would not automatically happen; there would have to be a way to guarantee that breeders responded to long-term public interest. A freer market in seeds might still lead to the replacement of traditional varieties in the vitally important centers of adaptation, Cary Fowler of the National Sharecroppers Fund, pointed out (*S:* 29). Opponents frequently called on the rhetoric of the market to condemn monopolistic tendencies they found in the bill. But they also appealed to a prior right: that of a society to the conditions of self-determination and to social (and ecological) stability. Kenneth Dahlberg, a political scientist, argued that because the legislation threatened "the basic genetic endowment" upon which human life depended, it was a "constitutional" issue (*H:* 131). The burden of proof should be on proponents to show that the rights they wished to create did not infringe on these prior rights (*H:* 135).

In practice this meant that opponents saw plant breeding as an issue of qualitative change, while proponents saw it in terms of the rate of progress. For opponents, because plant breeders could affect so greatly the range of seeds available and the properties those seeds had, they were necessarily making choices about public goods, which should be made by researchers in the public domain (*H:* 94; see also Busch and Lacy 1984b; Lipton and Longhurst 1989; Busch 1990; and Juma 1989). Proponents stressed that the pressure to improve seed made it essential to turn seed breeding over to private enterprise where the pressure would be most directly felt and lead to the most rapid improvement.

Figure 12.1 Some Substantive Differences

	Opponents	Proponents
Who should guide plant breeding?	public sector scientists	private sector scientists
What is the main injustice?	monopolization of seed stock	violation of breeders' rights
How will the PVPA affect breeding?	induce breeders to select for commercial interests only	provide incentives to breed more new varieties
What will certificate holders do with their certificates?	sell them to large seed companies allowing reduction in the range of available varieties.	produce certified seed; expand the range of available varieties.
Whose interests should be protected?	farmers	small seed companies
What policy would be dangerously destabilizing?	to pass the PVPA amendments	not to pass the PVPA amendments

These disputes about rights and utilities were not abstract debates about ethics or social philosophies. The two sides made concrete (and contrasting) claims of what would come from passage of the amendments. In both cases these appeared in the idiom of laissez-faire economics and both scenarios involved combinations of economic vectors and timely political actions. Both professed loyalty to the free market system and faith in the wisdom of the consumer, though in each case there were many qualifications. Both claimed to be concerned about the "little guy," be it the small seed company, the family farmer, or the third world peasant. They diverged on two central issues: First, how would certification affect the development of new varieties and the distribution of power within the seed industry? Second, how far could consumer wisdom be trusted? Were farmers' seed selections or consumers' food choices wise? If not, under what conditions would they be wise?

Let us focus first on new varieties. Would the amendments hasten development of new varieties? of varieties adapted to different kinds of farming? of varieties that led to greater ecological stability? Proponents said "yes"; opponents, "no." Proponents saw the legislation as finally making things normal, as fixing a problem that had become increasingly severe. "This is a free enterprise system in which each and every farmer and each and every seed company . . . can be actively involved in research," insisted Schlei of the USDA (*S:* 10). But without protection for nonhybrid plants, breeders had been reluctant to develop them; in effect they were being prevented from exercising their rights. Great pressure for improvement—both in terms of social need and unrealized genetic potential—had accumulated and could be released only by providing incentives.

In the proponents' view this pressure was felt most acutely by breeders working privately or as small firms. Anyone could apply for a certificate, they reiterated. With the protection of certification, even small firms could seek out market niches and develop products to fill them. Patents would finally give them a means to withstand the marketing dominance of larger firms; a wider range of seeds would come onto the market. Farmers would get better crops, consumers better food; forced by consumer demand, seed firms would compete to expedite progress as rapidly as possible.

Opponents saw the bill not as a return to the normal but as a great public gift that would in practice go only to a few large corporations. While proponents focused on who could apply, opponents focused on what would be done with the certificates granted; and where proponents saw an opening of opportunity, opponents saw a narrowing. Possession of a monopoly over a plant variety would make small seed companies a tempting target for takeover by larger firms. As the market became dominated by a few firms, farmers would be left with fewer varieties from which to choose.

The issue of what certificate holders would do with them was a touchy one, for pervading the hearings was the assumption that policy must uphold the "little guy" in competition with the giant and the powerful. Thus proponents felt a need to argue that "the little old farmer" who was here being given "the opportunity . . . to recoup the venture capital which goes into . . . research" would not sell that right (*S:* 12). They admitted that certificate owners could, in principle, dispose of them as they pleased, but they insisted that any concentration that had occurred in the seed industry since 1970 had been due to general business trends, not seed patenting laws. Opponents assumed that if independent breeders did obtain certificates (a rare event, they thought) they would sell them to the big firms. Indeed, they went so far as to suggest that what made the bill attractive to small firms was that it created a new salable asset—the certificate, not the seed.

By taking over small seed firms or their certified varieties, larger firms might gain monopoly control over seed stock for a particular crop, opponents feared. Especially if these were agrichemical firms, they might use their control to extend the industrial paradigm in agriculture. It was not necessary to imagine a conspiracy to drive up farmers' costs (though some opponents hinted that such was the case) but only to suggest that these firms were unlikely to mount breeding programs that would move away from dependence on chemical fertilizers and pesticides. Proponents were shocked by this suggestion—"preposterous," one called it. They appealed to economic law: "prices are a product of the marketplace" (*H*: 108). The reason that seed prices could not be fixed or the market dominated by unduly chemically dependent seeds was that farmers would not put up with it. It was inconceivable or irrational that a practice should prevail that was costlier than available alternatives.

Underlying these contrasting visions of where economic forces would take us were differing conceptions of human nature. Witnesses assumed they knew what choices free agents would make and also under what circumstances these agents were truly free. Each side portrayed its position as not only consistent with the behavior of innovative breeders but also with the behavior of farmers and consumers as purchasers of the products of breeding.

At issue with regard to farmers' seed selection was how to view farmers who planted traditional varieties (mostly third world farmers). Each side accused the other of taking the farmer for a fool. Charles Lewis argued it was unrealistic (and unfair) to think that third world farmers planted inferior traditional varieties as a matter of free choice; that they, unlike western farmers, were willing to sacrifice productivity

> as a contribution toward preventing some unpredictable and in any case rare event [e.g., a massive pestilence]. I would hate to argue with a Third World farmer or with an advanced world farmer that he should plant inferior seed and make a low production and low income as a public contribution toward the future good of the Earth (*H*: 203).

It was environmental imperialism to presume that third world farmers had different and less selfish priorities than western farmers, Lewis was suggesting, and unethical to deny them the full range of choices. Until a full range of seed was available to them, such farmers could not be presumed to have made a free choice; their agriculture would continue to be denied the rationalization of the market, and their economies would remain museum displays of outdated practices, dependent on the charity of the industrial world.

No one thought farmers should or would buy "inferior varieties" (though there was disagreement about whether inferiority was mainly a matter of yield), but opponents saw the restriction of free choice as lying chiefly in the seed industry's power over what varieties would be sold. In their view to imagine any farmer willingly participating in an overly concentrated seed market, where the decision about what to sell and how much to sell it for was not significantly influenced by buyers, was to cast that farmer as a fool. Whether farmers were income maximizers was not the issue; with certification would come fewer varieties and less power for farmers to affect prices and maximize their income.

A related disagreement arose with regard to whether food buyers were free choosers. Proponents defended the claim that agriculture must concentrate on a few high-yield varieties, by arguing that this was what consumers demanded. As George Sprague, Professor of Plant Breeding and Genetics at the University of Illinois, put it:

> The housewife demands uniformity. She wants to have the produce she picks up at the market of uniform maturity and uniform size, color, and so on. The commercial processor, the wheat miller, for example, needs to have wheat of uniform quality in order to satisfy demands for production of bread, for the production of flour for cakes and cookies (*H:* 85).

Opponents accepted this description of current consumer demand. But they saw it as a problem to be solved rather than a condition to be tolerated. Doyle of the Environmental Policy Center quoted Dr. Erna Bennett of the FAO:

> in a society where economics are dominated by market considerations it is inevitable that these will eventually eclipse all others in determining what is grown and offered for sale. What matters most in such circumstances is salability—shape, color, size, yield (insofar as it affects price and therefore marketability), popularity with the consumer and such factors (*H:* 232-3).

Failure to take into account nutrition or long-term issues such as resistance to pests and diseases meant that consumers were not really acting freely, for they were not sufficiently informed. At issue then, was not so much the appropriateness of the market ideal, but the question of what constituted a truly free market. To determine this both sides relied on assumptions about the principles on which consumers would (and should) chose among various goods, and each side saw some economic actions as necessary and therefore unproblematic and others as capricious and irrational, having no claim to be taken as true demand, and hence rightly subject to change.

Managing the Risks of Genetic Vulnerability

A great deal of testimony dealt with genetic vulnerability—the idea that through crop breeding the genetic base of the cultivars on which modern agriculture depends was being changed in a way that made them increasingly vulnerable to outbreaks of pests and disease epidemics. Other factors were acknowledged to enhance vulnerability: the practice of monoculture, the reliance on a few high-yield strains for each crop, and demands for uniform produce.

While there were a few scoffers among the bill's proponents—Don Dunner of the American Patent Law Association referred to the fear of genetic vulnerability as a "sky is falling" argument (*S*: 51)—both sides found the prospect of massive pestilence in genetically uniform crops a real one and both agreed that inherited resistance was the most important defense against pestilence. Memory of the Southern corn blight of 1970-1 was fresh. A fungus had mutated quickly and unpredictably; virtually the entire corn crop was susceptible to it and its rampages were only stopped (after it had destroyed 15 percent of the crop) by fortuitous weather and quick introduction of resistant varieties (NRC 1972, Doyle 1985). Even the stolid National Academy of Sciences had been much struck, concluding that "most major crops are impressively uniform genetically and impressively vulnerable" (NRC 1972: 1, 286ff).

Witnesses disagreed sharply about what the lessons of the corn blight were and how in general to respond to the genetic vulnerability problem. Proponents believed that we had to organize our resources, develop them further, and ready ourselves to mobilize them quickly. In practice this meant developing better capabilities for warehousing and classifying varieties, relying on breeding programs (including private sector programs) to develop more resistant stocks, and relying on the adaptability of a well-capitalized breeding industry to respond to new kinds of pests. By necessitating a program for characterizing and storing varieties, certification would contribute to this effort. Opponents responded that the resources needed to resist pestilence existed already in the wild or in traditional multivariety agriculture. Natural selection would keep pace with mutating pests; we would find the genetic resources needed to combat pestilence in wild regions and traditional agriculture, but only if we preserved such habitats. Some added that were we to go further and incorporate natural diversity in our own agriculture, the likelihood of serious pestilence would diminish greatly. But because it encouraged replacement of traditional varieties with genetically risky high-yield varieties, certification put us even more at risk.[1]

These scenarios and accompanying images of responsible action involved the full range of ideology. But most striking were contrasting conceptions of what nature was and would do (see Figure 12.2). Opponents of the bill saw nature as a beneficent legacy, a relatively stable accumulation of a long series of biological experiments. To be sure, evolution continued in this "storehouse" of germplasm but it would be dangerous to allow "tampering with the composition of that storehouse so as to restrict the range of its operation and evolution" (*H*: 234). In this view, nature had achieved something special: it had evolved to the degree where it gave us a "gift of wealth," the diverse and adaptable ecosystems that sustained us. "Our task," according to Doyle, was to "understand that diversity and use it constructively" (*H*: 233). Yet we were far from such understanding; the "symbiosis" between humans and plants was one "that we do not yet fully understand from the side of the plant." Daniel Smith recognized "a crisis in germplasm resources" due to destruction of the "complex web of gene pools and ecosystems of the tropics" (*H*: 237). Dahlberg spoke of a "basic genetic endowment" (*S*: 131).

These nouns—"gift," "resources," and "endowment"—recall notions developed in earlier chapters of nature as a benevolent and well made whole, an entity that has produced the human beings who can now maintain or disrupt it. Here nature is treasured; a "gift" is not to be squandered but kept lovingly out of respect for the giver; an endowment is to be preserved for its yield. The gift from nature was of far greater value than the superficial "pseudovarieties" that plant breeders were producing and patenting, varieties that according to Nabhan would quickly lose their uniqueness in cultivation (*H*: 65). Prudent farming meant recognizing and sustaining this diversity. Preserving natural habitats, searching for wild relatives, retaining traditional varieties, growing more varieties, and diversifying varieties within regions were all called for.

Figure 12.2 Contrasting Assessments of Strategies for the Preservation of Rare Germplasm

	Ft. Collins Storage Facility	Nature
Proponents	safe harbor	unstable, inaccessible, vulnerable
Opponents	vulnerable, incomplete, artificial	safe harbor

Supporters saw nature as a collection of processes, not an entity. The outcome of these was neutral at best; it exhibited no stability, reflected no benevolence or wisdom. The view was most fully developed by Caldwell and Charles Lewis, former Executive Secretary of the National Plant Genetics Resources Board, the U.S. agency responsible for overseeing germplasm collection and storage. Caldwell and Lewis agreed that genetic diversity was immensely important: traditional varieties and wild races did contain genes useful in breeding for resistance. But they did not see either the cultivation of traditional varieties or the preservation of natural habitat as a viable way, either socially or biologically, of dealing with the vulnerability problem.

In their view of nature, natural selection was destroyer as well as creator. They argued that the germ pool was constantly in flux. It might be true that in the vastness of nature there were some wild genes that had been of great value in a species' ability to withstand unusual conditions (including pestilences) in the past. But these genes reflected not a gift, not a proof of natural wisdom, but only a truism of natural selection: that a variety still endured showed only that it had the genetic material to survive so far. But survival in the past did not guarantee survival in the future. The difficulty of protecting crops against all manner of natural threats would continue undiminished. Pests evolved so unpredictably that even traditional varieties with lots of resistant genes were not necessarily immune. The struggle went on and on: "as we evolve new varieties, so have the pests evolved. So you have got a counterbalance battle going on. And if we do not continue to develop new pest resistant varieties, then sooner or later the farmer is going to be out of business, because this is his main defense against pests" (*S*: 36-7). The very artificiality of agriculture as they saw it—cultivating specialized varieties of nonnative plants and endeavoring to protect them from pests, which frequently were also introduced species—meant that "natural farming" was an illusion. They did not consider the possibility (nor did opponents press the point) of an agriculture in which a range of less specialized varieties were cultivated and in which natural selection was allowed to operate; alternatives to monoculture were not seen as viable. "The only solution to this problem [of genetic vulnerability]," summarized Sprague, "is the availability of a large number of varieties among which the grower can make his selection. This means, then, that the solution to this problem is intensification of the plant breeding and development process" (*H*: 85).

Because agriculture was essentially artificial, artificial selection (and protection) of varieties appeared to be the only viable response to the crop vulnerability problem. For proponents the "storehouse" in which the genetic treasure was guarded was the USDA facility at Ft. Collins,

Colorado, the world's largest collection of seeds, all organized by computerized catalogue, and ready for use in an emergency. At Fort Collins all would be protected, Congress mandated it; in nature by contrast, observed Lewis, "there is no fixed responsibility and a terrible record of losing material as the years go by" (S: 45, 44, 55).

Drawing on these contrasting conceptions of nature, adversaries branded one another as irresponsible. Opponents implied that scientists who favored the bill were deficient in ecological awareness; while Lewis represented opponents as opposing the course of science in general and plant breeding in particular (S: 44).

What Shall Be Done? What Can Be Done?

Ostensibly, the motive for the hearings was to take public action. A change was being considered; it remedied some problems, it might engender others.   Yet witnesses (and questioners) came with very different notions of the realm of action. Much of the attention focused not on what to do, but on what kind of problem was at hand and on what could be done—in effect, on possibility rather than policy.

Several attitudes were manifest in this discussion. Some argued, in effect, that the amendments were too minor to think of them in terms of significant options. The issues themselves were so trivial that, do what we would, a significant change was highly unlikely. Others saw the issues in the bills as significant, but held that our options were few; policy was needed to make law consistent with reality. Still others held that the bill embodied major matters, that a great range of possibility existed, but that the stakes were so high that it was much better to withhold changes until there was a stronger signal on what direction to take.   Finally, some saw an unacceptable present as making change imperative and saw the means as secondary. By and large, proponents held the first two attitudes, opponents the second two (see Figure 12.3). Let us take these in order.

The view that the matters under consideration were insignificant was held by de la Garza. Adding six plants to a list and making certificates valid for a further year were nothing more than fine tuning the original act. No great policy issues were at stake, he insisted to witness after witness. At the end of the hearings he expressed his frustration and exhaustion with the hearings to opponents Jack Doyle and Daniel Smith. The hearings were legislative, not oversight hearings, yet they had turned into oversight:

Figure 12.3 Alternative Views of the Historical Significance of the PVPA
              Amendments

1. *The matter is not really significant; nothing greatly important is at stake.*
2. *This is a significant moment, but there are no degrees of freedom, no viable alternatives.*
3. *These matters are too fraught with import to take any action without more information on impacts.*
4. *These matters are enormously significant, action is mandatory, the present is intolerable.*

It has turned into research. It is a very simple hearing on H.R. 999. Very
few have addressed H.R. 999. They have hit the act [i.e., the original
PVPA of 1970]. They have hit patents. They have hit [the] basic philosophy of patenting microorganisms. We have hit everything but the simple
H.R. 999 that says let's add tomatoes, cucumbers, okra, bell pepper, carrots
and one other. Nobody has addressed that part of it (249).

In de la Garza's view, his subcommittee was simply the wrong place to
raise the issues that had been raised, issues ultimately of the broad
direction of world agricultural policy and perhaps even of human
survival. He admitted these were important issues, but denied that they
arose in the bill or in the mandate of the committee. Thus de la Garza
emphasized the distance between a context so enormous as to be nearly
all-encompassing and a bill so minor as very nearly to be inconsequential.
Smith was frustrated too, but for precisely the opposite reasons:

we don't feel it is getting nearly the broad look that this whole matter [of
patenting of life forms] should be getting. . . .
   We recognize the proponents of this legislation and their feelings that
there are bigger forces at work here than just plant patenting. There are
some fundamental questions. We wonder who is addressing those, and
where? (H: 240)

The second view, that the issues were great but the options few, was
most vividly expressed in the Senate hearings by Barbara Schlei, one of
the USDA officials supporting the bill. Schlei described a juggernaut of
progress that would wipe out traditional agriculture whether we wished
it or not:

we cannot expect or want our farmers to return to low-yield varieties. . .

Automobiles have replaced the horse and buggy, electricity has replaced candles, tractors have replaced the plow. I think that some of us may wake up in the morning and wish that we could return to Walden Pond. But I cannot reverse that process. I do not think we want to reverse that process. We cannot return to planting low-yield native varieties. Our agriculture is the most productive sector of our economy, and the most productive agriculture in the world.

And to keep it the most productive, to face the needs of millions of people in this world to be fed, we have to continue our research base. We have to continue . . . so that we can increase the food supply (*S:* 9-10).

The matters about which opponents were testifying were, she assured Stewart, "of great concern," but we simply could not have things the way we might want them (10).

Unlike other defenders of modern agriculture, Schlei did not celebrate the technological adventure but instead appealed to a gritty realism. She contrasted the constraining force of history, a history of technical changes that came to pass seemingly without anyone's intent, with the openness of hope and the prospect of returning to Walden; and she concluded that history ruled, Walden was lost to the past. She admitted and even invited doubt—"I do not think we want to reverse that process"—yet reinforced the status quo with a claim of powerlessness—"I cannot reverse that process." With Walden beyond reach we can only "continue" and "continue" to meet what Schlei presents as the demands of our existence. Having realized that we can't have what we wanted, we are called at the end of the passage to recognize imperatives and do what we must: "We have to . . . increase the food supply." This is powerful rhetoric; in empathizing with those alienated by the rush of progress and in representing herself as facing necessity with dignity, Schlei removes the questions brought before the committee from the realm of political choice to that of acknowledging reality.

One can divide opponents between those who saw the bill as threatening to the margins they represented and those who saw the hearings as an occasion to insist on the unacceptability of modern agriculture. Katharine Anderson of Gardens for All, a nonprofit organization promoting home gardening, Ginger Nelson of the National Center for Appropriate Technology, and M. Woodrow Wilson of the National Farmers Union exemplified the former perspective. Their concern was that the implications of the original act were so vast and uncertain that it was wisest simply to wait and to study before taking any further action (*H:* 93, 98, 102). Anderson explained that she was present not so much to inform as to be informed. Could one be sure that plant patenting would not increase seed prices, she wondered; there was a prima facie case that it would (*S:* 60). Because no studies had been done, the question was in fact a criticism. Nelson put the view clearly: "This bill should not be passed. A number of studies should be done before any action is taken. There needs to be a response to the questions raised during testimony . . . before we chart a course which includes issues of such magnitude" (*H:* 94).

Such a view contrasts sharply with Schlei's depiction of a technological juggernaut. For Schlei, the bill, though minor, was important if we were to keep up with the rapidly moving present; it represented a legislative acknowledgment of the dynamism of the real world. For Nelson and Anderson, there was great danger in the present and corresponding need for great caution. While the present was a critical watershed, no technological imperative compelled us to choose immediately; we could wait for knowledge and understanding and then "chart" the most prudent "course." This was not blanket conservatism, nor policy paralysis. Rather in its appeal to deliberateness, it was a rejection of both the technological determinism and the insignificance arguments of proponents.

The final view was the most dramatic. The bill perpetuated the assumptions, goals, and institutions of conventional capital-intensive agriculture, asserted Dahlberg and Doyle. For them, the hearings were an occasion to expose that agriculture as unwise and unjust. For them further study was a minimally acceptable response (*H:* 133), for the crisis was upon us. We were at a "crucial juncture," "drifting" without a "comprehensive plan" toward a "second revolution in modern agriculture" that would place resources irrevocably in the hands of the giant corporations, according to Doyle (*H:* 227, 231-2). Dahlberg was the only witness who claimed to speak from a global perspective. He described the causes and consequences of desertification and deforestation, of shifts of population to cities in the third world, of overreliance on nonrenewable energy, and of the rapid concentration of land holdings. He depicted a world in crisis: "when one looks at the increasing rates at which species are being lost, . . . we find that this is a fundamental threat to the evolutionary potential of mankind." We had made "on the whim of the moment" dangerous changes that threatened "the fabric of society" (*H:* 131). Our response had been minimal; we were moving in the right direction only "a little bit" and "slowly and rather painfully" (*H:* 133).

For Dahlberg and Doyle, critics who saw a clear need to move society in new directions, the hearings became a forum of political action, not simply an opportunity to deliberate about action. To testify was physically to oppose the vectors that were taking us into a dangerous, possibly catastrophic future. It was not enough to pause and study before passing certification legislation; we had to try with all our might to put the brakes on in time and overcome the massive inertia of agricultural institutions. However trivial the issues, the hearings were an opportunity to divert the juggernaut, one which had to be seized lest the bill pass and add a little velocity to the descent into disaster.

What we have reviewed here is the explosion of a minimal legislative adjustment into a passionate drama about, as the Midwest Catholic Bishops put it, "the control . . . of life itself" (*S:* 34). The disagreement stretched far beyond defense of vested interests or conventional partisanship. Because the confrontation took place in public space, it was conducted in general terms—virtue, rights, truth, justice, necessity, wisdom. Restricted to ten-minute speeches, witnesses had no opportunity to present integrated reflections on these matters, even had they been fully aware of the sensibilities they were appealing to.

It is hard to avoid the conclusion that the hearings accomplished little: the ongoing disagreement about their proper scope, the reiteration of contradictory assertions by opposing witnesses, the failure of questioners to follow up witnesses' statements—these suggest that the foundation for legislation was no more satisfactory at the end of the hearings than at the beginning. Could the hearings have been more successful? If by that we mean could they have resolved differences between proponents and opponents, the answer is probably "no." But if we try to imagine a process in which facts were ascertained more completely, evaluations stated more fully, rationales made more explicit, and trains of questions more effective, the answer is "quite possibly"—with adequate preparation for the depth of disagreement and some structure for the comprehension of that disagreement, in short, with some way of coming to grips with the ideologies that made it impossible to confine the hearings to six new vegetables and one more year.

Notes

1. For more recent discussion of ethical and policy issues of germplasm conservation see Busch, Lacy, and Burkhardt 1988, and Raeburn 1990.

13

Research Agendas as Social Action

Much criticism of modern agriculture focuses on agricultural research. The perspective of researchers has been too narrow, it is claimed: agricultural research has produced manifold social and environmental consequences both unintended and undesirable. We must in future come up with new conceptions of agricultural research or some new "paradigm" or better way of ensuring accountability (Busch and Lacy 1983b: 41-2, 201; Strange 1984a: 120; Saponara 1988). Such criticisms ask us to learn from the past, but while it is easy enough to look back and see what went wrong, it is not at all clear what sort of process of research agenda setting can guarantee better results in the future (Thompson and Stout 1991). It would be wrong to think that agricultural researchers of the past never thought about the social or ethical implications of their work. Indeed, much of the criticism is aimed not so much at their innocence of these implications as at the particular groups and values researchers did serve—the wrong ones, according to critics. Yet even when broad values are shared troublesome questions remain of whether (and how) explicit values can guide research and produce desired changes, through a research agenda. The history of agricultural research is full of instances where the intentions of the researchers themselves backfired. If we are to commit ourselves to research that will produce a more equitable and environmentally stable agriculture (a goal all parties would seem to endorse) can we do better?

In this chapter we suggest ways in which looking at current research agendas as ideological (in the nonpejorative sense considered earlier) can help administrators anticipate the consequences of research. We begin with two questions: what is agricultural research? and what is a research agenda? We then focus on three agendas: *Agricultural Technology Until 2030: Prospects, Priorities, and Policies* by Glenn L. Johnson and Sylvan H. Wittwer, a 1984 report issued by the Agricultural

Experiment Station of Michigan State University; *Research Agenda for the Transition to a Regenerative Food System* by Richard R. Harwood, J. Patrick Madden and Medard Gabel, a 1983 publication of the Rodale Cornucopia Project; and *Beyond the Green Revolution,* a 1979 book by Kenneth Dahlberg. These agendas are ideological: they assess what is, what is necessary, and what is "the path to be followed" to a desired future (Dahlberg 1979: 214). Although these authors reflect the range of ideologies considered earlier, they set forth from similar assessments of the capabilities of agricultural science and the problems it must address. They differ greatly, however, on what social agenda should underlie agricultural research and on how knowledge relates to social change.

What Is Agricultural Research?

Before we can think about the problems of setting a research agenda it is necessary to explore the question of what such an agenda can do. Can the course of discovery and development be directed by a research agenda? Some will say "no," that research must be "nature-driven"—we can only search and revise our search by what we find.[1] If this is true, intentions don't much matter. Research may well have given us our present, but we could not have predicted it and have no better grounds for anticipating the future. Discovery of the molecular structure of the gene, for example, promises to revolutionize agriculture in ways that could not have been predicted fifty years earlier. In such a perspective, a long term master plan is little more than a gimmick for winning generous support: reality dictates what can be discovered; each discovery dictates the next on the agenda. Many people in basic research take such a view (Johnson and Wittwer 1984: 12).

Such a view of research contrasts starkly with much of what goes on in agricultural research. Agricultural research was created to serve the public. It is looked to by champions of conventional agriculture to develop the technologies needed to compete in international markets, and by critics for blueprints of an equitable and ecologically stable agriculture. Matters are too urgent for unguided science, both sides agree; we need a deliberate response, a research agenda that takes account of social goals and attacks social problems.

"Agricultural research" is thus an ambiguous concept (see Latour 1987). It has the peculiar features on the one hand of drawing us to a particular and less problematic future and on the other of reflecting that we must go where nature leads. People may speak of agricultural research both as an instrument of purposive social action (closed-ended) and at the same time as autonomous—we must let it unfold as it will in response to nature (open-ended) (see Figure 13.1).

Research agendas trade on this ambiguity. They are promoted as representing both natural truths and social goals; surely the strongest claim an advocate of a particular agenda can make is that one's agenda and one's social goals are what reality dictates (e.g., Dahlberg 1979: 222). Essentially, all the works we consider here do this; however much they may urge the pursuit of basic research, they also assume that basic research will bring us an expected future. Likewise, each holds its own agenda to be superior precisely because it reflects the host of problems reality has left us with, though each differs on what these are.

We do not take this ambiguity of "agricultural research" as sinister. We see it in part as a reaction to the struggle of numerous competing groups (including scientists themselves) to control what goes on in the agricultural university, and in part as a result of insufficient probing of such concepts as "research," "development," "applied science," and "technology." [2]

Even in the early years of public agricultural research in the mid nineteenth century, there was conflict between German-trained scientists wishing to make glorious careers in basic science and farmers who wanted certified fertilizers or better insecticides (Rosenberg 1976). More recently public research institutions have been called upon to serve the very different and often conflicting interests of consumers, commodity groups, agribusinesses, environmentalists, and proprietors of large and small farms. In many cases they have done so, but have been able to reconcile their multiple missions only within the loosest of frameworks. A recent OTA report complained, for example, that rather than articulating clear goals, public agricultural researchers in the U.S. operated under hopelessly vague mission statements, such as: "to provide an ample supply of nutritious foods for the consumer at reasonable cost with a fair return to the farmer within an agricultural system sustainable in perpetuity." Such statements papered over political conflict, but provided no useful guidance for research administrators (U.S. Congress OTA 1981: 133-4).

Figure 13.1   Open- vs Closed-Ended Research

	Expected Outcome	*Role of Research*
Closed-ended	known	to meet goals, obtain ends
Open-ended	unknown	to discover nature

But agricultural research agendas don't just suffer from the need to placate diverse clients. There is also great ambivalence about what research is.

Consider three widely held presuppositions about research. First, a common scheme distinguishes research from social action: researchers find things out; a separate decision is then made on what action to take. Second, the results of good research are usually expected to produce knowledge that is integrated and consistent. One may think this because one thinks nature (which research discloses) is itself integrated and consistent (Hacking 1986: 143-4). It would be quite unacceptable, for example, to find laws of chemistry incompatible with laws of physics. Finally, it is common to distinguish research as discovery, from creation or invention. Discovery is the uncovering of eternal truths; invention the realization of designs created by the mind.

None of these presuppositions fits all the disparate research activity that goes on in agricultural research institutions. There research may include studies of genetics, immunology or climatology, which involve acquisition of natural knowledge for its own sake (though in fields relevant to agriculture) (Busch 1984b: 300). It will also include develop-ment projects, such as attempts to make irrigation or feed conversion more efficient. For agricultural economists or environmental toxicolo-gists, research may entail advising policymakers. For experts in packag-ing, marketing, or animal science and crop science, research may include contributing to the marketing and production efforts of particular groups. So broad is the range of research activity that we may be able to say only that agricultural research is simply "what agricultural researchers do to fulfill the research obligations their universities or experiment stations place on them."

Given a strict reading of the common presuppositions, most of the "research" activity would have to be seen as something else, e.g., as social engineering, social action, policymaking, participation in agricultural markets, industrial development, public education, or even social criticism. The presuppositions do not hold true for most of this activity. It will be hard, for example, to separate discovery from application: even in basic research, agricultural universities operate on the assumption that research ought to contribute to application, and even basic research projects are often undertaken with the needs of particular client groups in mind (Thompson and Stout 1991: 6ff). As for consistency and integration, there is no guarantee that research results will be compatible: research undertaken at one end of campus may serve interests antithetical to those pursued at the other end. Finally, many development oriented projects are not only dealing with what is, but are trying to impose visions of what infrastructure, institutions, or the economic environment ought to be. In agriculture then, research agendas are often agendas for social change, sometimes for fairly significant change. But the wanted changes are presented, from each of many different viewpoints, as the obvious, natural, or inevitable course of science.[3]

How Do Research Agendas Differ?

We use "research agenda" in a loose sense here. The works we survey below conform to no common format and are written for different audiences in very different circumstances. They are in no sense parallel responses. Johnson and Wittwer, and Dahlberg write at much greater length than the Cornucopia authors; Johnson and Wittwer and the Cornucopia authors both provide lists of projects, while Dahlberg makes a plea for human survival. To compare these texts item by item, to point out that Dahlberg says nothing about research on root diseases, for example, would be naive and unfair. Moreover, even if one were to do this one would find fewer differences than one might expect: the three agendas agree substantially about what needs to be researched (cf Dahlberg 1979: 4). They all see a crisis in modern industrial agriculture, recognize that present methods are unsustainable, are concerned with social and environmental externalities, and recognize in new technologies (particularly biotechnology and computerization) the prospect of radically different agricultural institutions. Nor are there overt differences on methodology: all the authors call for interdisciplinary research and for the study of agricultural systems as well as individual components of systems. Indeed, it would be possible to select passages from each of the works that would sound like what one would expect to find in one of the others. Perhaps this is not surprising: what were once marginal issues—the fragility of the environment, the importance of quality of life, the nonrenewability of many agricultural inputs—have become well accepted.

What Makes a Good Solution?

The key ideological differences in the agendas are evident when we ask two sorts of questions of them. First, what do the authors regard as the characteristics of satisfactory solutions? Although the three agendas have similar lists of problems, they reflect quite different views of what things it is acceptable to change to solve a problem. Conflicting views of the issue of economic concentration in U.S. agriculture offer a simple example. Many people recognize low farm income as a problem. They differ not over this gloss of the problem, but in their judgments of what can or should be changed to solve it. Need we maintain a certain number of farmers? (In that case price supports may be necessary.) Or is the number of farms to be allowed to fall, raising average income for the remaining farms?

Another example is the matter of pesticides, appearance, and consumer choice, an issue that we saw earlier. Johnson and Wittwer, reflecting a conventional point of view, write that products "that differ appreciably in color, taste, texture, general appearance or storage quality, or that yield less, are not likely to be accepted" (1984: 17). By contrast, Harwood and Madden call for "educational and marketing programs for consumers, growers, and institutions regarding [USDA] #2 grade produce" and for efforts to "revise grading standards to reflect nutritional status, product end uses, maturity and eating quality, rather than simply external appearances" (1983: 21). For Johnson and Wittwer it would be misplaced effort to try to change existing consumer preferences—they limit the agronomic alternatives we can consider. For Harwood and Madden, on the other hand, current consumer preferences are part of a larger problem; we have to change buying habits to achieve ecologically sustainable practices (see Figure 13.2).

A final example comes also from our agendas: to offset the vulnerability to pestilence caused by using single strains of cultivars, Johnson and Wittwer suggest "multiline composites," the intermixing of different varieties of a single cultivar that are uniform in "height, maturity, grain type, quality and yield," but have differing and complementary capacities for disease resistance (17). For many alternative agriculturalists, the proper response to the same problem is intercropping and mixed farming (cf Harwood et al. 1983: 4). Both stress genetic diversity; they differ in the framework within which the problem must be solved. Wittwer and Johnson solve the problem within the dominant framework of monoculture; the crop—a grain crop—will be suitable for mechanical harvesting, for distribution and sale as a commodity. For alternative agriculturalists

Figure 13.2   How Conflicting Solutions Arise

Problem: Consumers tend not to buy the blemished fruit from pesticide-free farming		
	Factor to be Held Constant	*Solution*
Johnson and Wittwer	consumer preference	use pesticides in farming
Harwood and Madden	pesticide-free farming	change consumer preferences

this kind of farming is itself a problem; to solve it as Wittwer and Johnson do will only perpetuate an unsustainable and unsatisfactory practice.

Each agenda then, presumes a framework of criteria of acceptable solutions that is implicit, undefended, and reflects the ideology of its author(s). For Johnson and Wittwer, increasing productivity is the key problem; agriculture must continue to expand its capacity to produce. For Harwood and Madden, agriculture needs to sustain communities that produce whole and healthy people; current farming systems are problematic to the degree they prevent that from happening. In both cases, the solution criteria privilege some conditions as "natural" or obviously desirable—they can't be or shouldn't be changed. Yet such criteria are necessary; without them one would not know when a problem was solved and would even have trouble recognizing what the problem is (Schwarz and Thompson 1990: 33ff). In these agendas key differences lie in solution criteria, but because they are left implicit it is often difficult to sort out how the authors differ.

How Does Research Constrain Results?

The agendas also differ with regard to the open-endedness of particular research projects, a point we raised earlier (see Figure 13.1). Some projects are clearly intended to create something, a process, institution, or capability. Others seem genuinely open-ended; it is not clear what the political and economic significance of the research will be, and the results would seem to lend themselves equally well to all sorts of purposes. But most of the projects are not clearly of either type; even though they may seem open-ended, they are often relevant only to particular agricultural institutions and within particular assessments of possibility and propriety.

In keeping with the rhetoric of discovery and research, all the agendas make a virtue of open-endedness and open-mindedness. They are much more closed-ended than they appear, however: even empirical investigations of natural phenomena are presented as a means of moving us toward particular social goals. One way this happens is through omission. The authors omit the language of hypothesis and refrain from stating how an implicit hypothesis might turn out to be false. Areas of research are presented either in terms of empirical discovery—e.g., Gabel's call to "determine social and economic costs of continuation of present-day losses of farmland and topsoil" (Harwood et al. 1983: 33)—or development—e.g., "wider gene pools that simulate early populations need to be created" (Johnson and Wittwer 1984: 18).

In contrast, the language of hypothesis requires the recognition of conditions under which an hypothesis would be rejected. Consider, for example, Gabel's agendum above about topsoil loss. Suppose it were to be rephrased as "test the hypothesis that the current level of topsoil loss has significant social and economic costs." One would then be obliged to say under what conditions topsoil loss would have no significant social and economic costs and to consider what one might mean by significant costs. The original statement circumvents that process by begging the question. It assumes that significant costs exist, not "costs and benefits" as in some of Gabel's agenda items. Hypothesis language, by contrast, underscores the need to make epistemic and practical decisions in the process of agricultural research that are glossed over by the language of discovery and development.

Three Research Agendas

The three agendas we discuss correspond loosely to the ideologies we developed earlier. Johnson and Wittwer reflect a conventional productivist approach, though both have been open to critics of modern agriculture. Harwood, Madden, and Gabel have all had close ties to the research institute of the Rodale Press, publisher of *Organic Gardening* and *New Farm*. The Cornucopia Project, the Rodale sponsor of the agenda, shares radical humanist concerns for strong and stable communities. Dahlberg is concerned with the ecological viability of agriculture, much as was Walters, though he does not share Walters' populist outlook and often seems to speak for both wings of the alternative agriculture movement.

The Johnson-Wittwer Agenda

The Johnson-Wittwer report was initially part of a joint effort of Resources for the Future, the USDA, and the Joint Council of Food and Agricultural Sciences to determine what agricultural research would be needed to reach certain levels of production in the U.S. by 2030. The authors also addressed the delicate issues of coordinating the nation's agricultural research efforts and of responding constructively to critics of the agricultural research establishment. Focus on such "sensitive issues" and the desire for timely publication led to the report being published as a "special report" of the Michigan Agricultural Experiment Station (1984: iv).

Johnson and Wittwer outline the research necessary to permit a 100 percent increase in U.S. agricultural production capacity by 2030. This capacity need not all be used, they recognize, but it ought to be available to help balance a projected trade deficit, feed growing populations, and raise nutritional levels worldwide. Research will also be needed to compensate for likely shortages of land, water, and energy. The authors insist that agriculture must be made more efficient and stable, and it must provide safer and more nutritious food. It will also have to be carried on in such a way as "to prevent environmental pollution, contamination of the food chain and the development of undesirable rural, social and economic structures" (v-vi).

The report classifies research as either problem-solving (inter-disciplinary effort to resolve client problems), subject matter (multidisci-plinary research to develop bodies of information for decision makers concerned with broad problems such as energy use in agriculture), or disciplinary (concerned with problems internal to particular disciplines) (6). All these are important; a coordinated mix of them is necessary; and, in contrast with the 1972 Pound report, Johnson and Wittwer argue that disciplinary research must not be privileged at the expense of inter- or multidisciplinary inquiries that address practical problems (12).

Much of the report is a list of projects in the three categories. Problem-solving research is needed with regard to PBB contamination (particularly relevant to Michigan), soil erosion, farm labor, plant patenting, pesticide regulation, coordination of different types of production on the farm, farm computerization, the volatility of commod-ity prices, adaptation to new technologies, the ecological consequences of cash cropping, the wise use of public land, the volatility of international markets, and the improvement of crops suited to particular situations.

The items vary greatly. Some are time- and place-specific (PBB contamination), some are chronic and widely distributed (soil erosion), and some are permanent (public land use). In the case of some (price volatility, farm labor) it is not clear what is problematic about them—they are perceived as quite different problems by different social groups. Some are policy (and political) problems as much as research problems. What unites them is not some formal problem statement to which they all conform; instead they are situations in which something is seen to need fixing. Problem-solving research is clearly not open-ended, but Johnson and Wittwer say little about criteria of satisfactory solutions.

Subject matter research includes research on cropping systems, plant breeding, farm management (particularly farming systems), integrated pest management, horticulture, forestry, animal science, aquaculture, mechanization, and agricultural structures, among others. Particular problems frequently provide the warrant for this research too: plant breeding includes developing plant varieties suited to highly saline soils, for example (17). Disciplinary research—on, for example, photosynthesis, changes in atmospheric chemistry, biological nitrogen fixation, tissue culture and somatic cell fusion, growth regulators, animal reproduc-tion—is likewise often problem-oriented.   Sections on disciplinary research are organized according to their relevance for various sectors of agriculture: e.g., "Disciplinary Research Relevant for Plant Productivity" or "Relevant for Livestock" rather than research in biochemistry, genetics, etc.

The "problem-solving" - "subject matter" - "disciplinary" framework is not another way of representing the old hierarchy of pure and applied research, even if it retains some of that flavor. Johnson and Wittwer describe a landscape littered with real, practical problems—some of them in repairing errors of past and present, some in adjusting to present and future, some in taking advantage of opportunities. They write as administrators of a public research institution, as those who must assemble the resources and skills each problem requires. Some problems will be transitory and demand a temporary assemblage of talent from a variety of fields, others will demand long-term attention from a narrow range of specialists.

What remains ambiguous is whose problems these are; because Johnson and Wittwer are tied more closely than the other authors to public agricultural research, no specific normative viewpoint or social vision is evident in their list of problems and probable solutions. Doubtless many of the problems will be researcher-defined and will reflect existing research projects. It might seem appropriate to regard the agenda as opportunity-driven, and to conceive that many of the problems it lists became problematic only after a plausible alternative was evident (cf Busch and Lacy 1983b: 40-5). Yet they do assert strongly that an ethical determination, couched mainly in utilitarian terms, must be the means of administering this diverse range of projects.

The Cornucopia Agenda

By contrast with Johnson and Wittwer, the authors of the two alternative agendas are freer to link research to a specific social vision. They can ignore established current research and do so without apology; it is precisely their purpose to define alternatives.

The Cornucopia agenda is divided into two sections. The first, by Harwood and Madden, takes up "Food Production." Their goal is to "learn what we need to know to regenerate America's soils, farms, and farm communities, and thereby guarantee a sustainable and affordable supply of food for present and future generations of Americans" (Harwood et al. 1983: 2). At first glance this may seem indistinguishable from the goals of the Johnson-Wittwer agenda (cf Johnson and Wittwer

1984: 35). But where Johnson and Wittwer see an agricultural industry heavily involved in world commerce, Harwood and Madden favor a regionally based agriculture emphasizing settledness and sufficiency.

To achieve such sustainable agriculture the authors call for research on the on-farm cycling of various nutrient materials and for development of systems of husbandry that foster health and self-sufficiency as well. Much of the research is farm-centered; larger problems, such as worldwide climate change, do not figure heavily in the agenda, perhaps because it is assumed that stability on the small scale will produce stability on the large scale. Only in a few cases (the recycling of urban wastes, the feasibility of widespread urban gardening) is the concern with a larger social unit. Many items are also on the Johnson-Wittwer agenda: research on alternatives to chemical pesticides and herbicides, investigations of aquaculture, farm systems research, development of energy efficient farming and processing techniques, and biological nitrogen fixation. The Cornucopia agenda gives greater attention to soil chemistry and to research especially associated with alternative agriculture—studies of sophisticated intercropping schemes, for example, or the search for perennial grains to replace the annuals currently used. For the most part however, what makes the agenda "alternative" is not the items it includes, but the integration of those items with social goals and projected policies.

"From Farmgate to Homeplate," the second part of the Cornucopia agenda, is by Gabel alone and is concerned mainly with sociological and economic analyses of the existing food system and of alternatives. Even more clearly than Harwood and Madden, Gabel sees the present as a unique historical turning point, a time to recognize the absurdities of the present and adopt a new direction for the next few millennia. We are to begin with goal setting research: "U.S. agriculture and food system leaders, researchers, activists, [and] consumers" are to meet to determine goals for "U.S. agriculture, the U.S. food system, and the global food system." There is also to be normative inquiry into the proper goals to achieve "equitable regeneration of physical, biological, technological, economic, social and human resources" (26). We are then to reconcile the two lists and find a foundation for agricultural policy. These may seem to be open-ended questions, but much of the rest of Gabel's agenda presumes that they have been answered in a particular way.

Gabel goes on to call for economic and policy studies on such aspects of agriculture as subsidies, tax laws, employment policies, policies for management of energy, water, minerals, and soil, export-import policies, and so forth. There is also to be research on agricultural education, on coordination of agricultural policy, and on the relations of food to health.

Dahlberg's Agenda

Kenneth Dahlberg's research agenda appears in chapters on "New Approaches to the Future" and "Agricultural Strategies and Policies for the Future" in his book *Beyond the Green Revolution*. Dahlberg's is an agenda for survival, and secondarily for social justice. These larger stakes warrant more radical alterations. Where the Cornucopia writers want only to reorient agricultural research, he urges us to transcend the limits of the western scientific tradition to achieve a much grander science, a kind of ultimate geography integrating all physical, biological, and sociocultural systems (1979: 140, 218).

The problem with western science is its specialization. Interdisciplinary research, such as farming systems research, may help, but it is not the remedy. Instead, we must understand the big picture—the ecological and evolutionary constraints on our existence—and then deduce agricultures consistent with those constraints. Filling out the big picture demands far more ambitious global modeling projects. We shall have to discover the dynamics of the water cycles, the cycles of various fertilizer minerals, and the cycles that affect climate. The effects of human actions will need to be included in these models. Finally all these models will have to be integrated. Even leaving out the complex effects of human intervention, "the complexity of the . . . interactions is staggering" (Dahlberg 1979: 149). We know too little about these cycles, Dahlberg notes on several occasions; and what we know is so partial and Eurocentric as to be highly misleading.

In contrast with the vastness of this modeling project, Dahlberg has little to say about particular technologies or research items. These are of little immediate importance because the agenda of research must be to forestall the "collapse" of "society" or of "our agricultural system" (141, 165, 168). Once we have worked out the survival constraints, however, we can build agricultural systems consistent with them. For this reason Dahlberg opposes attempts to determine future research needs through extrapolation from current trends or imagined technical possibilities. Such forecasts privilege the highly unsatisfactory present and ignore the necessities imposed by "evolutionary constraints."

Agendas as Ideologies: Keeping Up with Reality

Is research a gathering of knowledge or an undertaking of social change? In making a research agenda it is hard to avoid this tension. It is hard to think about an agricultural project without thinking about what the knowledge involved will allow us to do and what it would be nice to be able to do. Agricultural research agendas are then, in an important sense, agendas of action. They are concerned after all with matters that require action: supplying food, balancing trade, ensuring food safety, protecting the environment, righting wrongs, strengthening rural communities. They are not simply alternative lists of projects we might take up but charts to the future.

Overall, the agendas considered here are closed-ended; each begins with a future we must obtain, the research is to get us there. Within the context of each particular agenda, individual items are equally closed-ended; the warrant for their inclusion is that the results they give will help to achieve the desired future. Indeed, in some cases, as we have seen, agenda items simply state components of a desired future. They are research problems only insofar as we are called to discover a way to realize them. If we turn around and ask what warrants the particular future the agenda describes, the answer is usually that that future is what the research will disclose—it will be the discovered future as well as an envisioned and desired future. As in the ideologies we considered in Part Two, a constructive circularity of argument helps us equate the future we desire with the future we will get.

Even when they seem to be focusing most intensively on the acquisition of knowledge, all of our authors are actively shaping a picture of the future and how it will come about. This is clearest in the Johnson-Wittwer agenda. Like the *Agriculture 2000* report considered earlier, their agenda is full of assertions of changes in agriculture that researchers will make. A paragraph on swine production, for example, consists of statements of future farming techniques intermingled with research problems. We are told that "swine production . . . *will emphasize* breeding at an earlier age" and that "most males *will either remain* intact or receive controlled-released testosterone implants to improve protein accumulation and energy utilization" (Johnson and Wittwer 1984: 28, ital ours). These are statements of what farm practice will be. Other statements more clearly refer to research projects, but they are projects which imply a narrow range of action. For example, the statement—"Ideal endocrine levels for ova implantation and embryonic survival and nutrient requirements will need to be determined"—rests on the assumption that ova implantation will become commonplace.

For many, what is described here is precisely what ought to be happening in public agricultural research. Researchers should be aware of where the openings for technical advance lie, hence of what farming will be like—that is what should guide their research and give it relevance. The farming that Johnson and Wittwer describe will, like present farming, be capital intensive, highly mechanized, and technically sophisticated. But what this reciprocal justification of future state and research program does not do is to offer a choice whether or not to select that future.

Because they are not writing from within the agricultural research establishment, the authors of the other agendas have at least the possibility of discriminating more sharply between what is and what could (and ought to) be. They do just that, calling on us to look around, discover our condition, and develop alternatives. But their agendas too are not open-ended; the research will uncover the truths that will warrant the particular social visions they advocate, and it will develop the means of realizing those visions.

The Cornucopia authors do try to retain a distinction between research and action. Harwood and Madden write:

> the following research agenda is a starting point for regenerating our entire agricultural system. If carried out, we will *learn* what we need to *know* to regenerate America's soils, farms, and farm communities, and thereby guarantee a sustainable and affordable supply of food for present and future generations of Americans (Harwood et al. 1983: 2, ital ours).

Here research is to establish an empirical basis for social action. Yet in promising that the agenda will work, i.e., will give us what we need for regenerative agriculture, the authors assume a particular description of regenerative agriculture. By imagining a future in which it has been achieved, they can identify the research that will bring about that future, just as the assumption that swine ova implantation will be important in future agriculture is taken to warrant the research that will make it possible. Of course, one could try to shape a research agenda around the questions "whether any regenerative agriculture is possible?" and "what kinds of regenerative agriculture are feasible?" (and what kinds of worlds they would give us). But in the Cornucopia agenda the feasibility or general nature of regenerative agriculture are not in question.

Any distinction one might want to make between knowledge gathering and implementation is undermined by the prominence in the agenda of demonstration projects and large-scale feasibility studies. Consider for example, the call for

> research . . . on ways to alter the present nationwide marketing channels to facilitate more local and regional food production and to anticipate the possible impact of those changes. Research and development efforts [are needed] . . . to determine economically feasible strategies for sustainable agricultural production in each region (17-18).

Like much conventional agricultural research, this project does not take place in the laboratory, but requires intervening in the real world to see what works. Its success is to be measured in the degree to which marketing channels are changed. Here, *research* and *development* are indissoluble partners, as they are implicitly in much of the rest of the agenda. We know that we have found economically feasible strategies only when we have implemented them; research is finished when real people act. There can be no demarcation between the end of the research and the beginning of the implementation; one can only trace the spread of new techniques from experimental trials to full application.

Dahlberg's agenda reflects a different, if no less acute, tension between discovery and action. As we have seen, Dahlberg's research agenda focuses on discovering the ecological systems equation for the whole earth. We will then know enough about our position and effect on the biosphere to understand the limitations it imposes. We will then know enough of peoples and cultures to adapt development to them. And we can then learn how to manage policy issues, even on a local level, in ways consistent with developmental and evolutionary constraints. In this future, the vast catalogue of knowledge, all indexed to its appropriate uses, will allow us to produce policies that are not just temporarily expedient, but developmentally and ecologically sound.

Here research is presented as being distinct from and preceding action. But the contrast between the magnitude of the knowledge we need and the enormity of the crisis we face raises a great dilemma. Ought we to act now in the absence of adequate and systematic knowledge, or wait with great risk until we know the right way to fix things? In Dahlberg's response, research and action are bound up in a network of mutual justification, just as in the other agendas. He argues that the wisest course is caution: until we discover our evolutionary limits we need to put the brakes on population growth and industrial despoliation. It might be argued that this political goal is what Dahlberg is really after; that playing up the threat of global calamity is a way of coercing us to accept a new order. Dahlberg does suggest that this interim solution will resemble the real solution that research will give us; both will promote diversity and flexibility, and enforce limits on human ambition. The vast research project is not, then, as open-ended as it seems. Whether or not we complete it may be moot; it calls on us to be content with modest permanence rather than chance a glorious suicide. Its very vastness symbolizes our own ignorance and supports the argument for prudence.

Research, Power, and Democracy

Many critics have objected that agricultural research has concentrated power and undermined democracy (see especially USDA 1981). In their view the hope that agricultural research would promote American values such as freedom, opportunity, self-reliance, and economic independence has not been fulfilled. Instead, they perceive much of the research as having weakened the market position of independent farmers and increased the gap between rich and poor (Thompson and Stout 1991). Alternative agriculturalists often present their programs as ways of redressing such injustice; both the Cornucopia authors (1-2) and Dahlberg (1979: 139) are explicitly concerned about this (as are Johnson and Wittwer also). Yet all three agendas treat issues of power and democracy ambiguously; and in that ambiguity one can see the same complaints arising all over again.

For Johnson and Wittwer the ambiguity comes in part from the multiple interests to which a public research agenda is subject. They do recognize that the benefits of some research fall mainly to certain firms or industries and strengthen them in a way that may not be good. In such cases counterbalancing research in the "public interest" will be necessary, they suggest, yet few of the listed projects are presented as warranting such research. They also recognize situations (as in consumer resistance to food additives or concerns about animal welfare) in which researchers will need to respond to public concerns, but they resist the claims of many new agenda groups for a voice in research policy.

What the agenda lacks then, is an explanation of how "the public interest" is to be determined and how particular projects are to be evaluated in terms of it. Beyond the specific goal of expanding production capacity, the authors do call on agricultural science to help establish "an agricultural social and economic structure that adds to rather than detracts from the quality of life for both rural and urban people" (v). But this goal is too vague to yield answers to many of the questions of central importance to critics: Whose interests is the research agenda to serve? What ethical criteria should be used to choose the right research path? To whom is agricultural research to be accountable?

To see how hard such questions are to answer, consider how Johnson and Wittwer handle the "problem" of farm laborers. They write: "Increased wage rates and reduced supplies of labor pose problems that are time-, place- and commodity-specific." One solution is "labor-saving equipment and supplies," an outcome of the work of "engineers, economists, plant breeders and animal scientists." Thus far, it may seem that the authors are no friends of agricultural laborers, whose wages are a "problem." But they also find research "problems" in "new institutional arrangements includ[ing] ... work environments, worker safety, housing, collective bargaining rights and education of the children of laborers." They conclude:

> Institutional aspects of the regulation of labor and the creation of labor-saving technologies in producing and harvesting fruits and vegetables are important. Some problems grow out of counterproductive institutional regulations, while others exist because there are no regulations. Institutional research could make a contribution to multidisciplinary efforts to solve labor problems in the fruit and vegetable industries (14).

The language leaves it unclear what is to be the relation between research and change. Are we simply being informed of areas of inquiry and disciplines that might undertake those inquiries? Or is this an agenda for agricultural researchers to help farm workers achieve their goals, or alternately to help ensure that the demands of farm workers will present fewer problems to the drive for greater productivity? That there can be "contribution[s]" to "solv[ing] labor problems" suggests that there is some unambiguous measure whether the situation is improving or worsening, but because the authors say nothing about whose problem this is, one has little basis for discerning what that measure might be.

How different is this neutral language from the language of conflict that we commonly use to talk about labor disputes, where participants see different problems (and sometimes see one another as "the problem"). It may be possible to achieve more or less durable compromises in farm labor disputes, possibly with aid from agricultural scientists; and while the compromises last the problem might be regarded as solved. But with respect to the autonomy of the parties in conflict, researcher-defined solutions are vastly different from such participant-defined solutions.

Throughout their report Johnson and Wittwer call for research that serves many different and conflicting interests. They point to the need for animal breeders to respond to public opposition to "use of feed additives, antibodies, pesticides, and herbicides for livestock and poultry" and on the same page predict that animal scientists will find ways of increasing feed utilization through the use of anabolic steroids (27). Even on the larger issue of whether agricultural research is servant or master, they write in one place that "improvements in human beings have become more fundamental and primary as our agricultural systems have developed," (5) and in another, that studies of "lifestyles and objectives" of rural people will have "to guide the creation of technology and its regulation and adoption" (38).

In part such ambiguity reflects the authors' recognition that agricultural research is, and presumably will continue to be asked to serve incompatible interests. But it also reflects a great and unusual faith in the possibility of solving agricultural problems through objective normative research linked to agricultural science. It is hoped that the "institutional research" on farm labor problems will provide more objective assessments of various "goods" and "bads" and thus lead to optimal solutions. Their agenda includes numerous normative research projects in which economists and rural sociologists (and presumably, agricultural ethicists) are engaged in making just such determinations.

Whatever its merits, this remains a perspective in which researchers retain power. It may replace the simplistic assumption that all products of research are inherently good with a more sophisticated normative analysis, but problems and solutions appear to remain researcher-defined; it is hard to escape the interpretation that a process of academic inquiry has come to substitute for democratic politics.[4]

That researchers retain (and even gain) power in the conventional productivist agenda may be expected, but the alternative agendas imply a concentration of power as well. Like many documents critical of modern agriculture, the Cornucopia agenda is concerned with empowerment—of consumers, gardeners, operators of small farms—all of whom are seen as victimized by contemporary food-producing institutions, policies, and research programs. Yet significant ambiguity over the relation between research and social change calls into question this commitment to empowerment.

Consider three ways that research might lead to the establishment of alternative forms of agriculture (see Figure 13.3). First, it might be undertaken to confirm that a particular form of agriculture is the best one. Such research would be truly open-ended, for it might well show instead that that form of agriculture was not the best. Second, the form of agriculture envisioned might be seen as the one free and responsible people would choose. With the basic principles already determined, the research would be undertaken to facilitate implementation, but it would still be somewhat open-ended. So long as they were consistent with the vision of agriculture these free and responsible people had opted for, whatever techniques the researchers found would be satisfactory. Third, it might be held that survival or justice requires a particular sort of agriculture, and hence that people must be persuaded to accept it. In this case the main function of research would be didactic. The necessity of the change would be accepted as given and the research would be undertaken to convince the rest of us of that necessity. Thus the research would be much more closed-ended; indeed it would be deeply embarrassing if it

Figure 13.3   Three Conceptions of the Relations of Agricultural Research to the Achievement of Alternative Agriculture

1) OPEN-ENDED: *Alternative Agriculture as an Hypothesis*	confirmed	accepted on empirical grounds	(research as hypothesis testing)
	falsified	rejected on empirical grounds	
2) SEMI-OPEN: *Alternative Agriculture as a Social Vision*	research facilitates		(research as development)
3) CLOSED-ENDED: *Alternative Agriculture as Imperative*	research ensures		(research as enforcement)

failed to indicate the need for the change or if demonstration projects failed to work.

Each of these brings with it a different conception of where power lies. In the first, research is relatively autonomous yet serves the public good; the decision to adopt a different form of agriculture is taken on its merits by citizens at large. In the second, research is less autonomous; it is guided directly by the public will and is concerned mainly with developing technical options that complement or support prior public choices. In the last, research is completely subservient to a prior determination of necessity; it is to show us what must be done and persuade us to do it. Here people are the primary subject of the research.

The latter two perspectives dominate in the Cornucopia agenda. Some agenda items are clearly matters of implementation. For example, there is a call to explore intercropping possibilities and to understand the relations of soil biota to the dynamics of soil chemistry, research projects that assume the social and agronomic viability of mixed organic farming (Harwood et al. 1983: 4, 8). In other cases, particularly in the section by Gabel, most of the research projects appear aimed at promoting particular social and technical policies. Take for example the call to "determine effects—social, economic, industrial, national—of increased regional self-reliance" (30). Ostensibly this is open-ended and empirical. But Gabel is clearly an advocate of increased regionalism, and since that conviction is not being held up to scrutiny, the expected function of this research will be to provide evidence to others of the viability of the Cornucopia alternative.

Tensions over the concentration of power are most acute in the Dahlberg agenda. Like many critics of modern agriculture, he too advocates regionally adapted agricultures that fit into stable and democratic polities (1979: 172). Such regionalism will involve "basic shifts in emphasis: from large irrigated farms to small peasant farms (often dry-land); from large-scale industrial agriculture to small-scale diverse farming; from national development needs to local needs" (183). The goal is "to make the marginal peasant more genuinely self-sufficient (in nonmonetary terms) while at the same time giving him a small cash surplus that can be used for education, etc." (212). It is a scenario of economic democratization, of liberation from the West, from one's own state, and from "bureaucrats, planners, and politicians" (184). These changes are conceived to be warranted by what the vast human-ecological research project discovers and by basic principles of prudence (such as promoting decentralization and diversity [167]).

But how will these changes happen? How are people to discover the new ecological truths? Who is responsible for ensuring that ecologically correct agriculture is put into practice? A political process will be necessary, Dahlberg writes:

> [The] ultimate tests, however, remain political—for people must be educated, persuaded and pressured to accept those changes in their life-style, work places, habits, and thinking that will move industrial society toward decentralization and greater diversity. The alternative would clearly appear to be some sort of major but largely unanticipated collapse, perhaps preceded by a period of increasing authoritarianism (1979: 167).

But as this passage suggests, because it is expert knowledge that justifies the changes, they must ultimately be imposed from the top down or disseminated from central authorities outward. Described in this way, the political process appears heavy-handed, though it may be necessary nevertheless.

Dahlberg does recognize practical problems of centralized knowledge in movements toward decentralization, though he does not acknowledge a contradiction. It is important, for example, to get ecological truths directly into the hands of local people; truths that fall into the hands of the powers that be are too apt to disappear (172-4). He sees a new role for the planner as a kind of knowledge broker "able to organize data and experts on location-specific processes and interactions" and to "encourage the local community to discuss and sort out its priorities regarding the area in question" (175).

Yet however much room Dahlberg leaves for local determination, the ultimate sources of power are ecological knowers who see all and filter out "stereotypes, selective perceptions, and universal assumptions" as well as the "distortions, lacunae, and value predilections" that have come with Western science (217).[5] What goes on in the expert-facilitated regional discussions and priority-setting sessions will be framed in terms experts provide and will be subject to demands of ecological sustainability. It is the experts who will be "mapping out strategies" (171) for local ecological technologies, and these will come as "packages," gifts from researchers to particular regions (184). The contrast with Berry could hardly be sharper: here the holistic expert sees all and knows better than the peasant. Recognizing these tensions, Dahlberg ends the book by calling for a new sort of "statesman," with "a clear awareness of the longer-term needs of a group . . . and the patience to pursue . . . [appropriate] strategies [of sustainability] through education, persuasion, and *genuine politics*" (224). Lincoln, the founding fathers, Gandhi, and Martin Luther King jr are examples. Still, the tension between democracy and technocracy does not go away. Indeed, it is a tension that may to some degree be inherent in the multiple roles of research in modern societies.

That agricultural research has great social impact seems beyond question. That its research agendas are, like it or not, agendas of social change or social action seems also beyond question, even if we hold that the social actions they lead to are neither coherent nor intended. Is there a way of making the consequences of research more responsive to what citizens want, of putting research under the control of a more *genuine* politics? That will depend in part on notions of what genuine politics includes and in part on how one comes to understand "research" and indeed "science."

The examples above suggest that present views on these matters are beholden to a certain research mystique in which research has come to subsume social direction and social change. Yet all the agendas, to one degree or another, recognize a need to reform agricultural research; to make it more truly subject to democratic controls. For some the co-option of politics by research will be proof of a deliberate and sinister reach for power by an already powerful clerisy. Certainly each of the research agendas attempts to draw rhetorical power from the supposition that the wanted future will also be the discovered future; even the more populist and antiestablishment ones do this. Yet that rhetorical move can be seen as less a will to power than a fascinating confusion, deeply seated in our culture, about what research is and how to reconcile technocracy and democracy. These views of agricultural research continue to lean on analogies to the physical sciences and on the supposition that there is really only one true research agenda, whose course, whether with regard to photosynthesis, soil tilth, or even rural social structure, is dictated by nature. Accepting this path is presented as the essence of rational choice; hence, rationality remains bounded and defined by expert knowledge, while "politics" retains an opposite stigma.

Some, like Johnson and Wittwer, who continue to see agricultural research as a social service, bring to light an additional problem. In their view agendas must be plotted in terms of ethical imperatives and social goals as well as natural constraints; and the ethical demands, values, and goals are to be determined by researchers—ethicists, agricultural economists, and rural sociologists. But because agricultural researchers have mistrusted ethics (and social science), their claims that the results of their research have been "good" have rested on no stronger foundation than the intuition of the researchers (Johnson and Wittwer 1984: 9-11). In effect they are left trying to be ethical without being able to say what it entails. But rather than turning questions of ethics over to the collective conscience of the polity, the Johnson-Wittwer agenda aims to empower a new group of experts (as do the other two agendas as well); the animal scientist's intuitive conviction that his or her research program is good is to be replaced by the ethicist's analysis and the sociologist's survey. The public will presents a valid claim only when it is translated through the filters of these disciplines.

Perhaps the hold of expert decision making on the contemporary political imagination is so great that these are the only choices open to us if we would address social goals and changes credibly. Yet it also seems plausible that the research mystique has clouded perceptions of political alternatives in the name of a "rationality" that has grown quite narrow, narrower even than the practices of rationality we engage in and sanction both in political arenas and in every day life. We doubt whether these politics can be eliminated, or wisely should be. We worry that inflated expectations of research—even research by ethicists and sociologists—will continue to frustrate us and deepen the symptoms of political malaise. Rather than continue to ask more of science than it can possibly deliver, perhaps we need to focus more on the process of politics to see how it may be strengthened. Perhaps both science and politics would benefit from such an equalization.

Notes

1. As Rosenberg points out, this view of agricultural science triumphed over its rivals during the progressive era, and agricultural researchers indeed came to insights of central biological importance. In Rosenberg 1976, see "Science, Technology, and Economic Growth: The Case of the Agricultural Experiment Station Scientist, 1875-1914," 153-72; "The Adams Act: Politics and the Cause of Scientific Research," 173-84; and "Science Pure and Science Applied: Two Studies in the Social Origin of Scientific Research," 185-95. See also Buttel 1986b; and the introduction to Busch 1981.

2. Among our authors, Johnson and Dahlberg address these matters elsewhere (Johnson 1986, 1990; Dahlberg 1986), but have not resolved the tensions we discuss below. Harwood is currently developing a more systematic framework that addresses biological and social diversity (1992); Dover and Talbot (1987) contribute on sustainable development. Numerous others are addressing the goals of agricultural research with an increased self-consciousness of these issues, e.g., in Thompson and Stout (1991) and in Johnson and Bonnen (1991).

3. The term "agenda" mirrors the ambiguity of "research." An agenda can be a nonprioritized list of items to be discussed (as in the "agenda" of a meeting), a prioritized list of items (as in "the first item on the agenda is . . ."), or an ascending series of goals to be achieved ("the next item on the agenda of my career is to . . ."). There are striking differences between these. In the first usage the term is open-ended. It obliges us only to consider each agenda item and presumes no particular decision or action. It brings with it images of deliberation and due process. In the last use the term refers to a goal-oriented activity, in which each item marks a step in the realization of a vision. It brings with it images of empowerment.

4. In fairness, Johnson and Wittwer are by no means uncritical advocates of the unrestricted expansion of agricultural science. They complain at length of current agenda-setting procedures that "appear to be a defense of the agricultural research bureaucracy rather than an earnest, honest effort to solve problems" (1984: 54).

5. The image of an all seeing eye is explicit. The process of contextual analysis Dahlberg outlines (of seeing things simultaneously in ecological, developmental, and policy contexts) is likened to "using the zoom lens on a movie camera" (1979: 218). One can focus at a particular level but always has the full reality available. But it is not an image unique to Dahlberg. Dahlberg is using the familiar "transparency of nature" language of Western science and arguing that we must equip ourselves to see more clearly.

14

Conclusion

What have we done? What have we found? Such questions are usually addressed in the conclusion of a scholarly study. Yet our answers—that we have constructed a detailed map of three major outlooks in contemporary agricultural controversy and that we have found it possible in a variety of ways to narrow the disjunctures between these views and thus help to bring them into fuller conversations with each other—will leave many traditional expectations unsatisfied. Where in each view should proponents be ready to give ground? Where should they reach out to forge alliances? Which presumptions must be abandoned? Which images are compatible, which might be synthesized? What policy recommendations are warranted? And above all, in what direction should agriculture go in the future? We have left all of these questions unanswered; we have interpreted the views but have offered no assessment or critique of them; we have discussed a wide range of policy issues but have offered no substantive policy recommendations; we have probed the different visions of the future of U.S. agriculture but have offered no vision of our own.

These are, of course, matters on which everyone would like to have answers. Thus our silence on them is all the more perplexing. Yet it is not for want of opinions of our own that we have chosen to remain silent. Rather, it is because there is something here that strikes us as vastly more important than seizing the opportunity to proclaim *our* opinions, mount *our* criticisms, advocate *our* solutions, or the like. That something is the impetus we hope this book will create to reshape, not policy outcomes necessarily, but policy processes, and to reshape them in ways that enhance the role of open discussion between disparate voices in making policy.

For us this book is the beginning of a larger enterprise—one larger than our opinions or anyone's opinions, larger than present policies or their alternatives, larger even than the question of who shall have the power to make policy decisions about agriculture. Within the U.S., not to mention many other places, discussions over the very basic matter of how we shall provide ourselves with food have passed during our lifetimes into an historical era of fracture and fragmentation, of disjuncture and ultimate impasse. In this process of historical change it is not only particular policies that have become threatened, but the very possibility of making effective policy democratically. The gaps between where different people are coming from have grown so large as to strain trust in democratic processes, particularly processes of open discussion. The frameworks that people use have grown so deeply divergent as to cast doubt on whether the other guy is even trying to be reasonable. In no one's definition of reason is it any longer to be expected that all concerned will respond positively to the Enlightenment call, "Come, let us reason together. . . ." Something different is needed to heal the rifts, and it must begin by respecting the divergences that have already occurred in people's ways of being reasonable, in their ideologies. Thus the larger enterprise is to bridge these differences without presuming on them, and not merely on paper, but in our political practices and especially in the policymaking process.

Many areas of policymaking, including agriculture, have long confronted a host of difficulties that impede both the exploration of alternatives and the justification of decisions. Much that policymakers would like to know is not known when it is needed, and may never be known. Yet conflicting assertions abound, and their makers come into the policy arena with impressive credentials from a disparate and daunting array of disciplines. Interests diverge and yet must be heard; new voices appear, with or without credentials, and clamor for attention. The public seems easily confused, often disheartened, even apathetic; yet in a democracy no more ultimate political recourse is available than to public opinion. To these difficulties are added, in the area of agriculture among others, a further fragmenting of interests, deeper divisions among voices, wider disparities of language, deflations of political capital, and shrinking resources for making policy. And as if this weren't already enough to sink the boat, to top it all off are added deeply rooted differences in the meanings, norms, and practices, of being a rational participant in the policymaking process.

The vastness of substantive and impassioned disagreement over matters of policy has contributed greatly to the fragmentation of policymaking in agriculture. Faced with a daunting range of groups who disagree so strongly about so much, agricultural policymakers and lobbyists alike find no alternative but to narrow the scope of policy deliberations to immediate issues that are small enough to be manageable and to limit influential voices to a few (Browne 1988). All too few among either participants or the managers of policymaking have endeavored to explore deep differences or to think very much about long-term policy. The result is agricultural policies that are doubtful in their consistency or coherence and hence rarely effective. Such policies are unlikely to be durable, unlikely to serve (or even be accountable to) any particular interest or institution, to embody any particular vision, to solve any particular problem, much less grapple successfully with the big problems, the long-term problems that are being raised. The best that can be hoped for, it would seem, is to muddle through with minimal and temporary compromises, hoping not to estrange so many parties that policymaking breaks down altogether.

We do not accept that this state of affairs is unavoidable. That the size of the issues and the extent of participation have hampered policymaking is clear, but this is not an inevitable fact of policymaking. On the contrary, that the presence of large issues and many voices has hampered rather than enhanced policymaking seems to us to reflect a pervasive, but unnecessary impatience with differences of outlook, and a widespread failure to confront those differences directly and openly.

It may also be the case that changes in the structure of policymaking in agriculture are in order. But one is not forced to think of large changes here—for example, of finding ways to neutralize the clout of special interests or imperious researchers, or bypass the harangue of ignorant citizens, or escape the perverse arbitrariness of legislators' concerns. Ultimately, reasons of equity or effectiveness may make large adjustments necessary, but as the language in which we describe them suggests, they almost invariably involve redistributing access and accordingly power—the partial disenfranchisement of some groups to the advantage of others. Without taking such radical measures, one can still find practical and minimally disruptive ways to enhance the level of dialogue in policymaking.[1]

It may also be argued that we need to increase the competence of agricultural policymaking to deal with big issues, that the items currently on the agenda are of such grave importance—economically, socially, culturally, morally, and ecologically—as to make business as usual unacceptable. Not all are convinced of the gravity of the present, but if the case is plausible at all it would seem prudent to take the condition of the policymaking process as seriously as we take the substance of policy on particular issues.

Considerations like these lead us to recognize a need to strengthen and extend the kind of skilled mediation that can facilitate communication and help those who disagree to recognize how it is that they manage to disagree. The sort of mediation we mean is not the formal sort that takes place between labor and management, or even between groups in conflict over the siting of highways, or powerlines, or the like. In the sorts of controversies we are dealing with, the issues are broader and more nebulous, and they are not necessarily joined at specific places and times. The wanted mediation will not likely be that of a go-between, carrying proposals and assessments between established groups in dire conflict, but ideally will appear in advance of sharp conflicts in the form of proactive efforts to shape policy openly with participation from many different voices. Among such efforts will be those of scholars to make changes—sometimes subtle changes—in the ways and circumstances in which they do research and write and teach about technological decision making in democratic societies. Satisfactory resolutions will be defined less in terms of the specific content of policies and more in terms of the confidence of the parties that their concerns truly have been taken into account and that they can trust in the direction of policy and in the people who will carry it out. The success of mediation will be correspondingly hard to assess. It will not appear as a strike averted or even as a declared public compromise, but in subtle changes in political discourse—such as greater sensitivity in the explication of policy proposals or greater flexibility in choices of language and rhetoric—leading perhaps to broader and more stable political coalitions and to policies that are durable, flexible, fair, and effective.

Will it work? Mediators cannot guarantee that their efforts will succeed. Nor can they ensure that participants will participate in good faith, or even participate at all. But they can make a plausible case why the effort to achieve better mutual understanding is appropriate. Our efforts in this book are predicated on such a case also, one that has involved four components:

1. an assessment of the current political climate in agriculture that finds it moving toward greater openness with increasing interest in creating an integrated policy more satisfactory to a wide range of groups.

2. an assessment of the distribution of political power among the various groups concerned with agricultural policy that finds that power too diffuse for any one group to successfully dominate policymaking.

3. a view of the prospects for improving mutual understanding that sees room to increase comprehension by making opposing sensibilities more visible to each other.

4. and an argument that the process of mediation itself, by facilitating better dialogue, can significantly improve the quality of political discussion—the strategies participants take, the immediate goals they seek to realize, their assessments of their success.

These four components are crucial to our case that attempts at mediating agricultural controversy are generally appropriate at this time: we have no doubt the aim is worthwhile; the conditions, we believe, are favorable; and we have found that there are workable ways to proceed.

With regard to the first point there seems more openness among those in power and less isolation and withdrawal among those alienated from power than was the case nine years ago, when we began this project. Proponents of alternative forms of agriculture have become less willing to remain on the political fringes and more concerned to find realistic ways to implement their ideas in mainstream agriculture. Within the agricultural establishment there seems to be an increasing desire to include a wider range of voices in policymaking and to include them in more meaningful ways (see Johnson and Bonnen 1991). While there remains bafflement on both sides about how it is possible to understand and accommodate the other, the movement on both sides seems real (if halting) and sincere (though qualified). Such interest and openness by no means guarantee the fruitfulness of mediation, but it would surely be hard for a mediative effort to get far without them.

With regard to the second point it can be argued that, in the present climate and from the perspectives of nearly all the participants, an engagement in mediation is more likely to achieve a happier result than a continued strategy of partisan opposition. So wide-ranging are the issues of agricultural controversy, so many are the groups taking stands on them, and so diverse the stands themselves that it will be very difficult for any single group to acquire sufficient power to effect its ends. There will have to be alliances and accommodations, and mediation can be a means to create them. This holds for the so-called "agricultural establishment" too: it is no longer as large as it once was or as united as it once was, and its political base has been much eroded.

With regard to the third point: a workable approach to mediation must possess some means to establish the legitimacy of the disagreement at hand. So long as participants have no way of understanding their opponents' positions as anything other than blatant error, moral badness, or utter confusion and ambiguity, they will have no reason to accommodate them. No mediation can succeed. Participants must either surrender or continue to assert their positions. The approach to mediation we anticipate here rests on the view that deep disagreement occurs (or can be most fruitfully comprehended) at the level of ideologies.

Construed narrowly as systems of belief, ideologies in their most basic assumptions may well be impervious to criticism. Yet by raising to consciousness contrasting sensibilities about human nature, knowledge, social order, and so on, we open up prospects for accommodation; one may then find reason to rethink some of one's beliefs or to search out new ways to act on them. Articulating contrasting sensibilities will not automatically make disagreements go away, but it will facilitate more satisfactory discussion of policy problems. It will allow all the participants to understand their disagreement as neither inaccessibly subjective nor as objective, but as publicly accessible and potentially workable through reasoned discussion.

With regard to the fourth point, we take the view that how people respond to policymaking problems—that is, with what degree of hope, sensitivity, and insight—depends very heavily on their assessments of what is possible in the way of communication. In so far as one perceives the roots of deep disagreements as subjective, even arbitrary, and beyond the realm of reasoned discussion, one is likely to find hope only in strategies of limitation, restriction, and exclusion: decide only on the basis of what is known with greatest certainty, restrict participation to those who know in those ways, and exclude those who cannot be trusted to be reasonable according to one's own standards of reasonableness. In short, when one is convinced that one cannot communicate meaningfully and effectively with serious challengers, the best strategy always seems to be to try to ensure an outcome that is rational by one's own lights. On the other hand, once deep disagreements are perceived as bridgeable there is room at least for mutual comprehension and, possibly, for accommodation. And if there is a reasonable hope of accommodation, then there is also reason to invest the effort to cultivate sensitivity and insight. With the hope, the motive, and the skill, we can approach problems in a spirit of exploration, limit defensiveness, and transmute differences into resources. So the act of participating in an ongoing mediation is itself likely to alter the kinds of political tactics taken and hence the quality and intensity of controversy.

The role of mediator is not the obvious one for scholar-authors to undertake; nor, we add, do we consider it the only proper, or even necessarily the most appropriate one. It is an altogether different mode of scholarship from what is usually practiced and requires a rethinking (and frequently an abandonment) of conventional goals, methods, and modes of expression and argument. One must continually be changing voices, sometimes bringing a searching skepticism to ideas one is genuinely attracted to, at other times giving the benefit of the doubt to arguments one still can't fathom (even after great efforts), but which one knows others take seriously. One must avoid many of the things authors regularly do, like awarding righteousness wreaths or rationality points to the characters or arguments in one's narrative; and yet one must try, sympathetically, to explain how it is that those characters can see themselves as righteous and take their arguments seriously. One must operate with different notions of authorship, recognizing that in effect one is in conversation with one's characters, who are to be thought of as real people rather than puppets and are to be assumed to have the right to respond.

For this reason, one cannot take refuge behind the bulwarks of method or theory; methods and theories may help to explicate, but they cannot defend or legitimate unless they are acceptable to all involved. One must be, in a sense, a methodological relativist, forbearing the temptation to judge, to make claims about what is true and real, and especially to privilege one's own views. Yet in another sense one must eschew relativism; one must be convinced that there are better and worse modes of policymaking and that it makes an important difference which ones are practiced. One must also pay great attention to the problem of reflexivity and try to recognize and control the situatedness of one's own activities as much as possible. One must accept that this situatedness is neither undesirable nor avoidable, that it can contribute to one's ends or away from their attainment, and that the artful writer or mediator must try to recognize and use it. If mediation is wanting in any of these respects, the effort to mediate will either not be worthwhile or it will not be workable.

What all this amounts to is a different kind of political role that scholars may take up—particularly, but not exclusively those who study issues of science, technology, and society—a move from being voices of authority to being facilitators of communication. Yet it will not be a wholly unfamiliar move, for always we have recognized ourselves as involved in a conversation, a kind of politics, within scholarly settings. But now we must recognize that the participants in the conversation are not just co-disciplinarians, or scholars, or even Americans; that the issues are not safely detached from the immediacies of action or the pressures of vested interests; and that the ways to address them are not confined by the norms of disciplinary inquiry—the conversation can no longer be kept within those narrow bounds. Yet far from abandoning rational inquiry, what is being suggested is an expansion of it, a willingness to make it take in much more territory, do more work, and deliver more—not by way of imperious declarations of truths, but through situated mediation, by bridging again and again the rifts and chasms that challenge democratic cohesion. These skills, this competence, and the will and character to use them wisely are needed not only in the area of agriculture but throughout the political management of science and technology, and in their interactions with individual lives, societies, cultures, and the nonhuman environment.

Notes

1. For example, the Dutch *sociale kaart* as described in Schwarz and Thompson (1990: 138ff); the Illinois roundtable discussions on 1990 agricultural legislation (Unnevehr 1990); the negotiation of pesticide controversies in Iowa (Maney and Hadwiger 1980); and the process of facilitated, interactive workshopping used in a series of nutrition forums at the University of Minnesota (Dieleman 1989a,b, 1990), and in the Biotechnology and Michigan Food and Agriculture Workshop (1991).

Select Bibliography

Abercrombie and Turner, 1978. Abercrombie, N., Turner, B.S. "The dominant ideology thesis." *Brit. J. Soc.* 29: 149-170.

Acres. *Acres U.S.A.: The Voice for Eco-Agriculture.* Raytown, MO.

Aiken, 1982. Aiken, William. "The Goals of Agriculture." In Haynes and Lanier, v. 1. pp. 29-54.

___, 1985. Aiken, William H. "On Evaluating Agricultural Research." In Dahlberg. pp. 31-41.

Albrecht, 1975. Albrecht, W.A. *The Albrecht Papers.* ed. Charles Walters jr. Acres U.S.A: Raytown, MO.

Albrecht and Murdock, 1990. Albrecht, Don E, Murdock, Steve H. *The Sociology of U.S. Agriculture: An Ecological Perspective.* Iowa State University Press: Ames, Iowa.

Allaby and Allen, 1974. Allaby, Michael, Allen, Floyd. *Robots behind the Plow: Modern Farming and the Need for an Organic Alternative.* Rodale: Emmaus, PA.

Altieri, 1987. Altieri, Miguel A. *Agroecology: The Scientific Basis of Alternative Agriculture.* Westview: Boulder, CO.

Apter, 1964. Apter, David E. ed. *Ideology and Discontent.* Free Press: New York.

ASA, 1980. American Society for Agronomy. *Agronomy: Solving Problems, Serving People.* ASA Special Publication #37: Madison.

Ashford, 1972. Ashford, Douglas. *Ideology and Participation.* Sage: Beverly Hills.

Baier, 1958. Baier, Kurt. *The Moral Point of View.* Cornell University Press: Ithaca, NY.

Baker, 1979. Baker, Frank H. "CAST and the Big T: Truth." Council for Agricultural Science and Technology, Paper #7.

___, 1980. Baker, Frank H. "Responsibilities and Rights of Food and Agricultural Scientists Served by CAST." Council for Agricultural Science and Technology. Paper #12.

Batie, 1990. Batie, Sandra. "Review of NRC, Alternative Agriculture." *Environment.* 32 (3, April): 25-28.

Battelle, 1983a. Battelle Memorial Institute, Columbus Division. *Agriculture 2000: A Look at the Future.* Sponsored by the Production Credit Associations. Battelle Press: Columbus, Ohio.

___, 1983b. Battelle Memorial Institute, Columbus Division. *The Future of United States Agriculture.* Battelle: Columbus, Ohio.

Baumgardt and Martin, 1991. Baumgardt, Bill R., Martin, Marshall A. eds. *Agricultural Biotechnology: Issues and Choices.* Purdue University Agricultural Experiment Station: West Lafayette, IN.

Bazelon, 1983. Bazelon, David L. "Governing Technology: Values, Choices, and Scientific Progress." In Iannone. pp. 132-140.

Bell, 1962. Bell, Daniel. *The End of Ideology: On the Exhaustion of Political Ideas in the Fifties.* Free Press: New York.

Bellah et al., 1985. Bellah, Robert, Madsen, Richard, Sullivan, William M., Swidler, Ann, Tipton, Steven M. *Habits of the Heart: Individualism and Commitment in American Life.* University of California Press: Berkeley.

Bender, 1984. Bender, Marty. "Industrial versus Biological Traction on the Farm." In Jackson, et al. pp. 87-105.

Berardi and Geisler, 1984. Berardi, Gigi M., Geisler, Charles C. eds. *The Social Consequences and Challenges of New Agricultural Technologies.* Westview: Boulder.

Berdahl, 1976. Berdahl, Robert M. "Prussian aristocracy and conservative ideology: A methodological examination." *Soc. Sci. Inform.* 15: 583-99.

Bernstein, 1976. Bernstein, Richard J. *The Restructuring of Social and Political Theory.* Harcourt, Brace, Jovanovich: New York.

_____, 1983. Bernstein, Richard J. *Beyond Objectivism and Relativism: Science, Hermeneutics and Praxis.* University of Pennsylvania Press: Philadelphia.

Berry, 1972. Berry, Wendell. "Discipline and Hope." In *A Continuous Harmony, Essays Cultural and Agricultural.* Harcourt Brace Jovanovich: New York.

_____, 1977. Berry, Wendell. *The Unsettling of America: Culture and Agriculture.* Sierra Club/Avon: Totowa, NJ.

_____, 1980. Berry, Wendell. "Abundant Reward of Reclaiming a 'Marginal' Farm." *Smithsonian.* 11 (5): 76-82.

_____, 1981. Berry, Wendell. *The Gift of Good Land: Further Essays Cultural and Agricultural.* North Point Press: San Francisco, CA.

_____, 1983. Berry, Wendell. "Restoring my Hillside Field." *The Country Journal* 10 (4): 45-9.

_____, 1984. Berry, Wendell. "Whose Head is the Farmer Using? Whose Head is Using the Farmer?" In Jackson, et al. pp. 19-30.

_____, 1987a. Berry, Wendell. *Home Economics.* North Point Press: San Francisco.

_____, 1987b. Berry, Wendell. "A Defense of the Family Farm". In Comstock 1987. pp. 347-60.

Bertrand, 1980. Bertrand, Anson R. "Public Responsibilities of the Scientists—A USDA Administrators View." In ASA pp. 7-15.

Besson and Vogtmann, 1978. Besson, J.M., Vogtmann, H. eds. *Toward a Sustainable Agriculture.* Verlag Wirz: Aarau, Switzerland.

Beus and Dunlap, 1990. Beus, Curtis E., Dunlap, Riley E. "Conventional versus Alternative Agriculture: The Paradigmatic Roots of the Debate." *Rural Sociology.* 55: 590-616

Bezdicek and Power, 1984. Bezdicek D., Power, J.F., eds. *Organic Farming: Current Technology and its Role in a Sustainable Agriculture.* ASA Special Publication #46, Proceedings of a Symposium, 29 Nov - 3 Dec. 1981. American Society of Agronomy/Crop Science Society of America: Madison.

Biotechnology and Michigan Food and Agriculture, 1991. "Biotechnology and Michigan Food and Agriculture: Issues and Recommendations." Report from the workshop April 3-4, Kellogg Center, Michigan State University. Department of Sociology, Michigan State University.

Black, 1978. Black, C.A. "Science in the Public Arena." In Pendleton. pp. 29-35.

Blatz, 1991. Blatz, Charles V., ed. *Ethics and Agriculture: An Anthology on Current Issues in World Context.* University of Idaho Press: Moscow.

Bluhm, 1975. Bluhm, William T. *Ideologies and Attitudes: Modern Political Culture.* Prentice Hall: Englewood Cliffs, N.J.

Bocock, 1986. Bocock, Robert. *Hegemony.* Tavistock: New York.

Boehlje, 1987. Boehlje, Michael. "Costs and Benefits of Family Farming." In Comstock. pp. 361-374.

Boeringa, 1980. Boeringa, R. ed. *Alternative Methods of Agriculture.* Elsevier: Amsterdam.

Bonanno, 1990. Bonanno, Alessandro ed. *Agrarian Policies & Agricultural Systems.* Westview: Boulder.

Bonnen, 1984. Bonnen, James T. "U.S. Agriculture, Instability, and National Political Institutions: The Shift from Representative to Participatory Democracy." In United States Agricultural Policy for 1985 and Beyond. Tucson: Department of Agricultural Economics, University of Arizona. pp. 53-83.

Boody, 1982. Boody, George M. "Sustainable Agricultural Futures: Preliminary Goals and Criteria to Assess Growth." In Haynes and Lanier, v. 2. pp. 597-612.

Boulding, 1963. Boulding, Kenneth E. "Agricultural Organizations and Policies: A Personal Evaluation." In Farm Goals in Conflict. pp. 156-166.

Bourdieu, 1977. Bourdieu, P.R. *Outline of a Theory of Practice.* Cambridge University Press: Cambridge.

Bradfield, 1981. Bradfield, Stillman. "Appropriate Methodology for Appropriate Technology." In Usherwood. pp. 23-33.

Bramwell, 1989. Bramwell, Anna. *Ecology in the 20th Century: A History.* Yale University Press: New Haven, CT.

Breimyer, 1980. Breimyer, H. "Economics of organic agriculture." Paper #35. University of Missouri, Columbia.

Brewster, 1970. Brewster, John M. *A Philosopher Among Economists: Selected Works of John M. Brewster.* Ed. by J. Patrick Madden and David E. Brewster. J.T. Murphy Co. Philadelphia, PA.

Brewster et al., 1983. Brewster, David E., Rasmussen, Wayne D., Youngberg, Garth. *Farms in Transition.* Iowa State University Press: Ames.

Brooks, 1982. Brooks, Joseph F. "Agriculture and Human Values: A Perspective on Black and Small Farmers/Landowners in the Rural Southeast." In Haynes and Lanier, v. 1. pp. 247-61.

Brown, 1981. Brown, Lester R. *Building a Sustainable Society.* W.W Norton: New York.

Browne, 1987. Browne, William P. "Bovine Growth Hormone and the Politics of Uncertainty: Fear and Loathing in a Transitional Agriculture." *Agriculture and Human Values.* 4: 75-80.

___, 1988. Browne, William P. *Private Interests, Public Policy, and American Agriculture.* University Press of Kansas: Lawrence, KS.

Browne and Lundgren, 1987. Browne, William P., Lundgren, Mark H. "Farmers Helping Farmers: Constituent Services and the Development of a Grassroots

Farm Lobby." *Agriculture and Human Values.* 4 (2 & 3, Spring, Summer): 11-28.

Buesching, 1978. Buesching, Don. "Review of Wendell Berry, The Unsettling of America." *Rural Sociology.* 43: 514-516.

___, 1979. Buesching, Don. "The Alternative Agriculture Movement: Origins, Development and Current Composition." *Ag. World.* 5: 1-8.

Busch, 1978. Busch, Lawrence. "On Understanding Understanding: Two Views of Communication." *Rural Sociology.* 43: 450-475.

___, 1980. Busch, Lawrence L. "Structure and Negotiation in the agricultural sciences." *Rural Sociology.* 45: 26-48.

___, 1981. Busch, Lawrence ed. *Science and Agricultural Development.* Allanheld, Osmun: Totowa, NJ.

___, 1982. Busch, Lawrence. "History, Negotiation, and Structure in Agricultural Research." *Urban Life.* 11: 368-384.

___, 1984a. Busch, Lawrence. "The Social Construction of the Natural World." Paper presented at a seminar on "Les Interfaces de la Connaissance des Milieux Naturels," at the École des Hautes Études en Sciences Sociales. Paris, France.

___, 1984b. Busch, Lawrence. "Science, Technology and Everyday Life." *Research in Rural Sociology and Development.* 1: 289-314.

___, 1989a. Busch, Lawrence. "Remembrance of Things Past (and Future): Plant Germplasm Conservation in France." Presented to annual meeting of the Rural Sociological Society, Seattle, WA, August.

___, 1989b. Busch, Lawrence. "Irony, Tragedy, and Temporality in Agricultural Systems, or How Values and Systems are Related." *Agriculture and Human Values.* 6: 4-11.

___, 1990. Busch, Lawrence. "Review of Lipton and Longhurst, New Seeds and Poor People." *Agriculture and Human Values.* 7: 112-114.

Busch and Lacy, 1981. Busch, Lawrence, Lacy, William B. "Sources of influence on problem choice in the agricultural sciences: the New Atlantis revisited." In Busch. pp. 114-127.

___, 1983a. Busch, Lawrence, Lacy, William B. "Information Flows in Research and Extension: An Alternative Perspective." *The Rural Sociologist.* 3 (2): 92-97.

___, 1983b. Busch, Lawrence, Lacy, William B. *Science, Agriculture, and the Politics of Research.* Westview Press: Boulder.

___, 1984a. Busch, Lawrence, Lacy, William B. eds. *Food Security in the United States.* Westview: Boulder, CO.

___, 1984b. Busch, Lawrence, Lacy, William B. "Sorghum Research and Human Values." *Agricultural Administration.* 15: 205-222.

___, 1986. Busch, Lawrence, Lacy, William B eds. *The Agricultural Scientific Enterprise: A System in Transition.* Westview: Boulder.

Busch, Lacy, and Burkhardt, 1988. Busch, Lawrence, Lacy, William B., Burkhardt, Jeffrey. "Culture and Care: Ethical and Policy Dimensions of Germplasm conservation." Presented at Beltsville Symposium XIII, Biotic Diversity and Germplasm Preservation -- Global Imperatives, May.

Busch and Sachs, 1977. Busch, Lawrence, Sachs, Carolyn. *The Social and Economic Organization of the Agricultural Sciences: A Preliminary Bibliography.* Lexington, KY.

Buttel, 1983. Buttel, Frederick H. "Beyond the Family Farm." In Summers. pp. 87-107.

___, 1986a. Buttel, Frederick H. "Agricultural Research and Farm Structural Change: Bovine Growth Hormone and Beyond." *Agriculture and Human Values.* 3 (4 Fall): 88-98.

___, 1986b. Buttel, Frederick. "Biotechnology and Agricultural Research Policy: Emergent Issues." In Dahlberg. pp. 312-347.

Buttel and Flinn, 1976. Buttel, Frederick H., Flinn, William L. "Sociopolitical Consequences of Agrarianism." *Rural Sociology.* 41: 473-483.

Buttel and Gillespie, 1988. Buttel, Frederick H., Gillespie, Gilbert W. "Agricultural Research and Development and the Appropriation of Progressive Symbols: Some Observations on the Politics of Ecological Agriculture." Bulletin No. 151. Department of Rural Sociology: Cornell University.

Buttel and Gontlev, 1982. Buttel, F. H., Gontlev, M. E. "Agricultural structure, agricultural policy and environmental quality: some observations on the context of agricultural research in North America." *Agriculture and Environment.* 7: 101-119.

Buttel and Larson, 1979. Buttel, Frederick H., Larson, Oscar W. "Farm science, structure and theory intensity: an ecological analysis of U.S. agriculture." *Rural Sociology.* 44: 471-488.

Buttel and Newby, 1980. Buttel, Frederick H., Newby, Howard eds. *The Rural Sociology of the Advanced Societies.* Allenheld, Osmun & Co.: New Jersey.

Buttel et al., 1982. Buttel, F.H., Larson, O.W., Harris, C.K., Powers, S. "Social class and agrarian political ideology: a note on determinants of political attitudes among full and part-time farmers." *Social Forces.* 61: 227-283.

___, 1986. Buttel, Frederick H., G.W Gillespie jr, R. Janke, B. Caldwell, M. Sarrantonio. "Reduced-Input Agricultural Systems: Rationale and Prospects." *Journal of Alternative Agriculture.* 1: 58-64.

Butz, 1976. Butz, Earl L. "An emerging market-oriented food and agricultural policy." *Public Administration Review.* 36: 137-142.

___, 1978. Butz, Earl L. "Review of Wendell Berry, The Unsettling of America." *Growth and Change.* 2: 52.

Callicott, 1990. Callicott, Baird J. "The Metaphysical Transition in Farming: From the Newtonian-Mechanical to the Eltonian-Ecological." *Journal of Agricultural Ethics.* 3: 36-49.

Campbell, 1985. Campbell, Brian L. "Uncertainty as Symbolic Action in Disputes Among Experts." *Social Studies of Science.* 15: 429-53.

Campbell, 1990. Campbell, James. "Personhood and the Land." *Agriculture and Human Values.* 7: 39-43.

CAST, 1972. Council for Agricultural Science and Technology. "Informing the Nonagricultural Public About Agricultural Science." Special Publication #2 (December).

___, 1990. "Alternative Agriculture, Scientists' Review." Special Publication #16. Council for Agricultural Science and Technology: Ames, Iowa.

Castle, 1978. Castle, Emery N. "Resource Allocation and Production Costs." In Gardner and Richardson. pp. 1-51.

___, 1985. Castle, Emery N. "World Hunger, U.S. Agriculture and Rural Resource Policy." Brigham Young University.

Catton and Dunlap, 1980. Catton, W.R. jr., Dunlap, Riley E. "A New Ecological Paradigm for Postexuberant Sociology." *American Behavioral Scientist.* 24: 15-47.

CED, 1962. Committee for Economic Development. *An Adaptive Program for Agriculture.* CED: New York.

Chan, 1963. Chan, Wing-Tsit, trans. *A Source Book in Chinese Philosophy.* Princeton University Press: Princeton, NJ.

Chibnik, 1987. Chibnik, Michael ed. *Farm Work and Fieldwork: American Agriculture in Anthropological Perspective.* Cornell University Press: Ithaca, NY.

Chinn and Navarro, 1984. Chinn, M. David, Navarro, Peter. "Capture and Ideology in American Farm Policy." *Rural Sociology.* 49: 517-529.

Chubin and Restivo, 1983. Chubin, Daryl E., Restivo, Sal. "The Mooting of Science Studies: Research Programmes and Science Policy." In Knorr-Cetina and Mulkay. pp. 53-83.

Cochrane, 1963. Cochrane, Willard W. "Beliefs and Values underlying Agricultural Policies and Programs." In *Farm Goals in Conflict.* pp. 50-63.

___, 1979. Cochrane, Willard W. *The Development of American Agriculture: A Historical Analysis.* University of Minnesota Press: Minneapolis, MN.

___, 1982. Cochrane, Willard W. "American Agricultural Development, Rural Values and Societal Valuation Conflicts." In Haynes and Lanier, v. 1. pp. 1-13.

Cohn, 1970. Cohn, Norman. *The Pursuit of the Millennium.* Oxford University Press: New York.

Coleman, 1982. Coleman, Eliot W. "Impediments to Adoption of an Ecological System of Agriculture." In Haynes and Lanier, v. 2. pp. 580-96.

Collin, 1985. Collin, Finn. *Theory and Understanding: A Critique of Interpretive Social Science.* Basil Blackwell: N.Y.

Colman and Elbert, 1984. Colman, Gould, Elbert, Sarah. "Farming Families: The Farm Needs Everyone." *Rural Sociology and Development.* 1: 61-78.

Comstock, 1987. Comstock, Gary ed. *Is there a Moral Obligation to Save the Family Farm?* Iowa State University Press: Ames.

Conner and Hessel, 1980. Conner, John T., Hessel, Dieter T. eds. *The Agricultural Mission of Churches and Land-Grant Universities: A Report of an Informal Consultation.* Iowa State University Press: Ames.

Constance et al., 1990. Constance, Douglas H., Gilles, Jere L., Heffernan, William D. "Agrarian Policies and Agricultural Systems in the United States." In Bonanno, 1990. pp. 9-75.

Converse, 1964. Converse, Philip. "The Nature of Belief Systems in Mass Publics." In Apter. pp. 206-261.

Conviser, 1982. Conviser, Richard. "Appropriate Agriculture." In Haynes and Lanier, v. 1. pp. 436-52.

Copeland, 1980. Copeland, W.R. "Ethical Dimensions of the Energy Debate: The Place of Equity." *Soundings.* 63: 159-77.

Cornucopia, 1981. Cornucopia Project. *The Empty Bread Basket: The Coming Challenge to America's Food Supply and What We Can Do About It*. Rodale Press: Emmaus, PA.

Cotgrove, 1982. Cotgrove, Steven F. *Catastrophe or Cornucopia: The Environment, Politics, and the Future*. John Wiley: New York.

Cotgrove and Duff, 1981. Cotgrove, S., Duff, A. "Environmentalism, values and social change." *Brit. J. Soc.* 32: 92-110.

Crosson, 1982. Crosson, Pierre R. ed. *The Cropland Crisis: Myth or Reality?* John Hopkins University Press for Resources for the Future: Baltimore.

Dahlberg, 1978. Dahlberg, Kenneth. "An Evaluation of Research Strategies for Developing Appropriate Agricultural Systems and Technologies." Revised Ms. of paper to International Studies Meeting.

___, 1979. Dahlberg, Kenneth A. *Beyond the Green Revolution*. Plenum: New York.

___, 1986. Dahlberg, Kenneth A. ed. *New Directions for Agriculture and Agricultural Research: neglected dimensions and emerging alternatives*. Rowman and Allanheld: Totowa, NJ.

___, 1991. Dahlberg, Kenneth. "Sustainable Agriculture: Fad or Harbinger." *Bioscience*. 41: 337-340.

Danbom, 1979. Danbom, David B. *The Resisted Revolution: Urban America and the Industrialization of Agriculture, 1900-1930*. Iowa State University Press: Ames.

Dasmann, 1973. Dasmann, R.F., Milton, J.P., Freeman, P.H. *Ecological Principles For Economic Development*. Wiley: London.

Davis and Hinshaw, 1957. Davis, John J., Hinshaw, Kenneth. *Farmer in a Business Suit*. Simon and Schuster: New York.

Day, 1978. Day, Boysie E. "The Morality of Agronomy." In Pendleton. pp. 19-27.

DeCrosta, 1980. DeCrosta, Anthony. "The Real Scoop on the Plant Patent Controversy". *Organic Gardening*. (May): 108-114.

DeJanvry, 1980. DeJanvry, Alain. "Social Differentiation in Agriculture and the Ideology of Neopopulism." In Buttel and Newby. pp. 155-168.

Delind, 1991. Delind, Laura. "Sustainable Agriculture in Michigan: Some Missing Dimensions." *Agriculture and Human Values*. 8: 38-45.

Dickson, 1974. Dickson, David. *The Politics of Alternative Technology*. Universe Books: New York.

Dieleman, 1989a. Dieleman, Linda S. ed. *Actions for Lowering Blood Cholesterol: Collaborative Strategies*. Proceedings of the Food, Agriculture, and Nutrition Forum I, June 3, 1988, University of Minnesota, St. Paul Campus; and report of the Educational Design Team. Sponsored by the Intercollegiate Nutrition Consortium, University of Minnesota.

___, 1989b. Dieleman, Linda S. ed. *Food Access: Exploring Issues and Affecting Changes*. Proceedings of the Food, Agriculture and Nutrition Forum III, July 12-13, 1989; and report of the Educational Design Team. University of Minnesota, Crookston Campus.

___, 1990. Dieleman, Linda S., Presider. "Food Talk: Managing Mixed Messages." Food, Agriculture and Nutrition Forum V, May 9, 1990, University of Minnesota, St. Paul Campus.

Dittberner, 1979. Dittberner, Job L. *The End of Ideology and American Social Thought*. UMI Research Press: Ann Arbor, MI.

Douglas, 1966. Douglas, Mary T. *Purity and Danger: An Analysis of the Concepts of Pollution and Taboo.* Praeger: New York.

___, 1982. Douglas, Mary T. "Environments at risk." In Barnes, B. and D. Edge eds. *Science in Context: Readings in the Sociology of Science.* MIT Press: Cambridge, MA.

Douglas and Wildavsky, 1982. Douglas, Mary, Wildavsky, Aaron. *Risk and Culture: An Essay on the Selection of Technical and Environmental Dangers.* University of California Press: Berkeley, CA.

Douglass, 1984. Douglass, G. K. ed. *Agricultural Sustainability in a Changing World Order.* Westview: Boulder.

Dover and Talbot, 1987. Dover, Michael, Talbot, Lee. *To Feed the Earth: Agro-Ecology for Sustainable Development.* World Resources Institute: Washington, D.C.

Doyle, 1985. Doyle, Jack. *Altered Harvest: Agriculture, Genetics, and the Fate of the World's Food Supply.* Viking: New York.

Dubos and Ward, 1972. Dubos, R.J., Ward, B. *Only One Earth.* W.W. Norton: New York.

Dumont, 1980. Dumont, Louis. *Homo Hierarchicus.* University of Chicago Press: Chicago.

Dundon, 1982. Dundon, Stanislaus J. "Hidden Obstacles to Creativity in Agricultural Science." In Haynes and Lanier, v. 2. pp. 836-868.

Dunlap and VanLiere, 1978. Dunlap, R.E., VanLiere, K.D. "The new environmental paradigm." *J. Environ. Educ.* 9: 10-19.

Dunlap, 1981. Dunlap, Thomas. *DDT: Scientists, Citizens, and Public Policy.* Princeton University Press: Princeton, NJ.

Dunn, 1979. Dunn, John. *Western Political Theory in the Face of the Future.* Cambridge University Press: New York.

Durrenberger, 1986. Durrenberger, E. Paul. "Notes on the Cultural-Historical Background to the Middlewestern Farm Crisis." *Culture and Agriculture.* 28: 15-17.

Ebenreck, 1982. Ebenreck, Sara. "A Partnership Farmland Ethic." In Haynes and Lanier, v. 1. pp. 453-61.

Edens, 1985. Edens, T. C., et.al, eds. *Sustainable Agriculture and Integrated Farming Systems: 1984 Conference Proceedings.* MSU Press: E. Lansing.

Edens and Haynes, 1982. Edens, Thomas C., Haynes, Dean L. "Closed system agriculture: resource constraints, management options, and design alternatives." *Annual Review of Phytopathology.* 20: 363-95.

Edens and Koenig, 1980. Edens, Thomas C., Koenig, Herman E. "Agroecosystem management in a resource-limited world." *BioScience.* 30: 697-701.

Edminnster, 1977. Edminnster, T.W. "Legislation: Does it control crop and soil science research?" In Pendleton. pp. 37-43.

Ekirch, 1973. Ekirch, Arthur jr. *Man and Nature in America.* University of Nebraska Press: Lincoln.

Elkana, 1981. Elkana, Yehuda. "A programmatic attempt at an anthropology of knowledge." In Mendelsohn and Elkana. pp. 1-77.

Encyclopaedia Britannica, 1984. "Ideology." 9:194-198.

EPA Journal, 1988. "Agriculture and the Environment." *EPA Journal.* 14.

Farm Goals in Conflict, 1963. Iowa State University Center for Agriculture and Economic Development. *Farm Goals in Conflict: Family Farm, Income, Security.* Iowa State University Press: Ames

Feyerabend, 1978. Feyerabend, Paul. *Science in a Free Society.* New Left Books: London.

Finer, 1972. Finer, S.E. "The Transmission of Benthamite Ideas 1820-1850." In Gillian Sutherland ed. *Studies in the Growth of Nineteenth Century Government.* Rowman and Littlefield: Totowa NJ. pp. 11-32.

Flanagan, 1972. Flanagan, Thomas. "Social Credit in Alberta: A Canadian 'Cargo Belt'" *Archives de Sociologie des Religions.* 34-48.

Foucault, 1980. Foucault, Michel. *Power/Knowledge: Selected Interviews and Other Writings 1972-1977.* Ed. by C. Gordon; Trans. by C. Gordon et al. Pantheon Books: NY.

Freudenberger, 1982. Freudenberger, C. Dean. "The Ethical Foundations of the Idea of Agricultural Sustainability." In Haynes and Lanier. v. 2, pp. 622-32.

___, 1990. Freudenberger, C. Dean. *Global Dust Bowl: Can We Stop the Destruction of the Land Before It's Too Late.* Augsburg Press: Minneapolis.

Friedland, 1982. Friedland, William. H. "The End of Rural Society and the Future of Rural Sociology." *Rural Sociology.* 47: 589-608.

Friedland and Barton, 1975. Friedland, W.H., Barton, A. *Destalking the Wily Tomato: A Case Study in the Social Consequences of Agricultural Research.* Davis, CA.

Fung, 1948. Fung, Yu-Lan. *A Short History of Chinese Philosophy.* Ed. Derek Bodde. The Free Press: New York.

Gamson, 1975. Gamson, William A. *The Strategy of Social Protest.* Dorsey Press: Homewood, IL.

Gardner and Richardson, 1978. Gardner, Bruce L., Richardson, James W. eds. *Consensus and Conflict in U.S. Agriculture: Perspectives from the National Farm Summit.* Texas A&M University Press: College Station, TX.

Geertz, 1964. Geertz, Clifford. "Ideology as a cultural system." In Apter. pp. 47-76.

___, 1979. Geertz, Clifford. "From the native's point of view: on the nature of anthropological understanding." In Rabinow and Sullivan.

Gendel et al., 1990. Gendel, Steven M, Kline, A. David, Warren, D. Michael, Yates, Faye eds. *Agricultural Bioethics: Implications of Agricultural Biotechnology.* Iowa State University Press: Ames.

Giddens, 1979. Giddens, Anthony. *Central Problems in Social Theory: Action, Structure and Contradiction in Social Analysis.* Macmillan: London.

___, 1984. Giddens, Anthony. *The Constitution of Society: Outline of the Theory of Structuration.* University of CA Press: Berkeley.

Gilbert and Mulkay, 1984. Gilbert, G. Nigel, Mulkay, Michael. *Opening Pandora's Box: A sociological analysis of scientists discourse.* University Press: Cambridge.

Glacken, 1967. Glacken, Clarence J. *Traces on the Rhodian Shore: Nature and Culture in Western Thought from Ancient Times to the End of the Eighteenth Century.* University of California Press: Berkeley.

Glaser and Strauss, 1976. Glaser, B., Strauss, A. *The Discovery of Grounded Theory.* Aldine: Chicago.

Gliessman, 1984. Gliessman, Stephen R. "An Agroecological Approach to Sustainable Agriculture". In Jackson, et al. pp. 160-171.

Goldschmidt, 1978. Goldschmidt, Walter. *As You Sow*. Allanheld Osmun: Montclair, NJ.

___, 1982. Goldschmidt, Walter, "Agricultural Production and the American Ethos." In Haynes and Lanier, v. 1. pp. 406-22.

Goodman, 1970. Goodman, Daniel. "Ideology and Ecological Irrationality." *Bioscience*. 20: 1247-1252.

Goodman, 1978. Goodman, Nelson. *Ways of World Making*. Hackett: Cambridge, MA.

Goodwyn, 1978. Goodwyn, Lawrence. *The Populist Moment: A Short History of the Agrarian Revolt in America*. Oxford University Press: Oxford.

Gouldner, 1976. Gouldner, Alvin W. *The Dialectic of Ideology and Technology*. The Seabury Press: New York.

Goulet, 1971. Goulet, Denis. *The Cruel Choice: A New Concept in the Theory of Development*. Atheneum: New York.

___, 1986. Goulet, Denis. "Three Rationalities in Development Decision-Making." *World Development*. 14: 301-17.

Gray, 1985. Gray, David B., Borden, R.J., Weigel, R.H. *Ecological Beliefs and Behaviors: Assessment and Change*. Greenwood Press: Westport, Conn.

Greeley, 1989. Greeley, Andrew. "Protestant and Catholic: Is the Analogical Imagination Extinct?" *American Sociological Review*. 54 (August): 485-502.

Gross, 1953. Gross, Bertram. *The Legislative Struggle: A Story in Social Combat* McGraw Hill: New York.

Grube, 1977. Grube, J.W., et al. "Behavior change following self-confrontation: A test of the value mediation hypothesis." *Journal of Personality and Social Psychology*. 35: 212-216.

Gulley, 1974. Gulley, James L. *Beliefs and Values in American Farming*. USDA Economic Research Service: Washington, D.C.

Habermas, 1979. Habermas, Jurgen. *Communication and the Evolution of Society*. Trans. by T. McCarthy. Beacon Press: Boston.

___, 1984. Habermas, Jurgen. *The Theory of Communicative Action: v. 1, Reason and the Rationalization of Society*. Trans. by T. McCarthy. Beacon Press: Boston, MA.

___, 1989. Habermas, Jurgen. *The Theory of Communicative Action: v. 2, Lifeworld and System: A Critique of Functionalist Reason*. Trans. by Thomas McCarthy. Beacon Press: Boston, MA.

Hacking, 1986. Hacking, Ian. "Culpable Ignorance of Interference Effects." In D. Maclean ed. *Values at Risk*. Rowman and Allenheld: Totowa, NJ. pp. 136-53.

Hadwiger, 1980. Hadwiger, Don F. "Achieving Agricultural Research Missions." In Conner and Hessel. pp. 39-56.

___, 1982. Hadwiger, Don F. *The Politics of Agricultural Research*. University of Nebraska Press: Lincoln.

Hadwiger and Browne, 1978. Hadwiger, Donald F., Browne, William F. eds. *The New Politics of Food*. Lexington Books: Lexington, MA.

Hadwiger and Talbot, 1965. Hadwiger, Don F., and Talbot, Ross B. *Pressures and Protests: The Kennedy Farm Program and the Wheat Referendum of 1963: A Case ,Study*. Chandler Publishing: San Francisco.

Halcrow, 1977. Halcrow, Harold G. *Food Policy for America* McGraw-Hill: New York.

Hamilton, 1963. Hamilton, W.E. "Goals and Values underlying Farm Bureau Policies." In *Farm Goals in Conflict.* pp 64-76.

Hamlin, 1992. Hamlin, Christopher. "Green Meanings: Thoughts about what Sustainable Agriculture might Sustain" *Science as Culture.* 13: 507-31.

Harris, 1979. Harris, Craig K., Powers, Sharon E., Buttel, Frederick H. "Myth and reality in organic farming: a profile of conventional and organic farmers in Michigan." Read at Rural Sociology Association Annual Meeting, August: Burlington, VT.

Harwood, 1981. Harwood, Richard R. "Agronomic and Economic considerations for Technology Acceptance." In Usherwood. pp. 35-47.

___, 1983. Harwood, Richard R. "International overview of regenerative agriculture." Presented to Tanzania Symposium on Regenerative Agriculture. Rodale Research Center: Kutztown, PA.

___, 1984. Harwood, Richard, R. "Organic Farming Research at the Rodale Research Center" In Bezdicek and Power. pp. 1-17.

___, 1992. Harwood, Richard R. "The Structure of Biological Diversity at the Agricultural, Environmental and Social Interface (an agricultural perspective)." Keynote address: Diversity in Food, Agriculture, Environment and Health. Michigan State University. June 4-7. East Lansing, Michigan.

Harwood et al., 1983. Harwood, Richard R., Madden, J. Patrick, Gabel, Medard. *Research Agenda for the Transition to a Regenerative Food System.* The Cornucopia Project: Emmaus, Pa.

Hayami and Ruttan, 1971. Hayami, Yujiro, Ruttan, V. *Agricultural Development.* J. Hopkins University Press: Baltimore.

Haynes and Lanier, 1982. Haynes, Richard, Lanier, Ray eds. *Agriculture, Change and Human Values: Proceedings of a Multidisciplinary Conference.* 2 Vols. Humanities and Agriculture Program: University of Florida.

Hays, 1975. Hays, Samuel P. *Conservation and the Gospel of Efficiency: The Progressive Conservation Movement 1890-1920.* Atheneum: New York.

___, 1987. Hays, Samuel P. in collaboration with Barbara D. Hays. *Beauty, Health, and Permanence: Environmental Politics in the United States, 1955-1985.* Cambridge University Press: Cambridge.

Heady, 1961. Heady, Earl O. ed. *Goals and Values in Agricultural Policy.* Iowa State University Press: Ames, Iowa.

Hesse, 1978. Hesse, Mary. "Theory and value in the social sciences." In Christopher Hookay and P. Pettit eds. *Action and Interpretation: Studies in the Philosophy of the Social Sciences.* Cambridge University Press: New York. pp. 1-16.

Hicks, 1979. Hicks, Jack. "Wendell Berry's Husband to the World: A Place on Earth". *American Literature.* 57:238-54.

Hightower, 1973. Hightower, Jim. *Hard Tomatoes, Hard Times.* Schenkman: Cambridge, MA.

Hildreth, 1982. Hildreth, R.J. "Normative Information, Positive Information and Policy in Agriculture." In Haynes and Lanier. pp. 55-66.

Hileman, 1990a.   Hileman, Bette. "Alternative Agriculture." *Chemical and Engineering News.* 68 (5, March): 26-40.

___, 1990b. Hileman, Bette. "Alternative Agriculture: Concept Finds Unexpected Support." *Chemical and Engineering News.* 68 (11, June): 4.

Hill, 1979a. Hill, S.B. "Ecology, ethics and feelings." In *The ReEvaluation of Existing Values and the Search for Absolute Values.* 2:1150. Int. Cult. Fdn.: N.Y. pp. 593-607.

___, 1979b. Hill, S.B. "Eco-agriculture: the way ahead?" *Agrologist.* 8: 9-11.

Hollander, 1986. Hollander, Rachelle D. "Values and Making Decisions about Agricultural Research." *Agriculture and Human Values.* 3 (3, Summer): 33-40.

Hollander and Aiken, 1985. Hollander, Rachelle, Aiken, William H. "Social Ethics, Agricultural Change, and Agricultural Research." Presented to AAAS Annual Meeting, May.

Huang, 1979. Huang, Shu-min. "Changing Taiwanese Peasants' Concept of Time: Its Impact on Agricultural Production." *Iowa State Journal of Research.* 54: 191-215.

Hutchcroft, 1982. Hutchcroft, Theodore. "What They Don't Know Can Hurt You." CAST Paper #14.

___, 1983. Hutchcroft, Theodore. "Responding to Media Cheap Shots: Observations on the CAST Experience." CAST Paper #16.

Iannone, 1987. Iannone, A. Pablo ed. *Contemporary Moral Controversies in Technology.* Oxford University Press: New York, NY.

Ihde, 1964. Ihde, A. *The Development of Modern Chemistry.* Harper and Row: New York, NY.

I'll Take my Stand, 1930. *I'll Take my Stand: the South and the Agrarian Tradition by Twelve Southerners.* Harper: New York.

Inglehart, 1977. Inglehart, R. *The Silent Revolution: Changing Values and Political Styles among Western Publics.* Princeton University Press: Princeton, NJ.

Jackson, 1984. Jackson, Dana. "The Sustainable Garden." In Jackson, et al. pp. 106-114.

Jackson, 1980. Jackson, Wes. *New Roots for Agriculture.* Friends of the Earth: San Francisco.

___, 1984. Jackson, Wes. "A Search for the Unifying Concept for Sustainable Agriculture." In Jackson, et al. pp. 208-29.

Jackson and Bender, 1984. Jackson, Wes, Bender, Marty. "Investigations into Perennial Polyculture." In Jackson, et al. pp. 183-94.

Jackson et al., 1984. Jackson, Wes, Berry, Wendell, Colman, Bruce eds. *Meeting the Expectations of the Land: Essays in Sustainable Agriculture and Stewardship.* North Point Press: San Francisco.

Jansen, 1975. Jansen, A.J. *Constructing Tomorrow's Agriculture.* Bull., 38:90. Agr. University, Wageningen: The Netherlands.

Jenny, 1984. Jenny, Hans. "The Making and Unmaking of a Fertile soil." In Jackson, et al. pp. 42-55.

Jeske, 1981. Jeske, Walter E. ed. *Economics, Ethics, Ecology: Roots of Productive Conservation.* Soil Conservation Society of America: Ankeny, Iowa.

Johnson, 1984. Johnson, Glenn L. "Academia Needs a New Covenant for Serving Agriculture." Mississippi Agriculture and Forestry Experiment Station Special Publication (July), Mississippi State University.

____, 1986. Johnson, Glenn L. *Research Methodology for Economists: Philosophy and Practice*. MacMillan: New York.

____, 1987. Johnson, Glenn L. "Roles for Social Scientists in Agricultural Policy." In Comstock. pp. 153-175.

____, 1990. Johnson, Glenn L. "Ethical Dilemmas Posed by Recent and Prospective Developments with Respect to Agricultural Research." *Agriculture and Human Values*. 7 (3-4, Summer-Fall): 23-35.

Johnson and Bonnen, 1991. Johnson, Glenn L., Bonnen, James T., with Darrell Fienup, C. Leroy Quance, and Neill Schaller, eds. *Social Science Agricultural Agendas and Strategies*. Michigan State University Press: East Lansing.

Johnson and Wittwer, 1984. Johnson, Glenn L., Wittwer, Sylvan H. "Agricultural Technology until 2030: Prospects, Priorities, and Policies." Agricultural Experiment Station Special Report #1 (July), Michigan State University: E. Lansing.

Juma, 1989. Juma, Calestous. *The Gene Hunters: Biotechnology and the Scramble for Seeds*. Princeton University Press: Princeton.

Kaufman, 1982. Kaufman, Maynard. "Visions for the Future of Agriculture." In Haynes and Lanier, v. 1. pp. 67-86.

Kiley-Worthington, 1981. Kiley-Worthington, M. "Ecological agriculture: what it is and how it works." *Agriculture and Environment*. 349-381.

Killingsworth and Palmer, 1992. Killingsworth, M. Jimmy, Palmer, Jacqueline S. *Ecospeak: Rhetoric and Environmental Politics in America*. Southern Illinois University Press: Carbondale and Edwardsville.

King, 1949. King, James E. *Science and Rationalism in the Government of Louis XIV, 1661-1683*. Johns Hopkins University Studies in Historical and Political Science, Series 66 #2. Johns Hopkins University Press: Baltimore.

Kluckhohn, 1951. Kluckhohn, C. "Values and value-orientations in the theory of action." In T. Parsons and E.A. Shils eds. *Toward a General Theory of Action*. Harvard University Press: Cambridge, MA.

Knorr, 1981. Knorr, Dietrich. *Sustainable Food Systems*. AVI Pub. Co.: Westport.

Knorr-Cetina, 1981. Knorr-Cetina, Karen D. *The Manufacture of Knowledge: An Essay on the Constructivist and Contextual Nature of Science*. Pergamon: New York.

Knorr-Cetina and Mulkay, 1983. Knorr-Cetina, Karen D., Mulkay, Michael eds. *Science Observed: Perspectives on the Social Study of Science*. Sage: Beverly Hills.

Kramer, 1977. Kramer, Mark. *Three Farms: Making Milk, Meat and Money from the American Soil*. Little, Brown: Boston.

Kramer, 1978. Kramer, John. "Agriculture's Role in Government Decisions." In Gardner and Richardson. pp. 204-41.

Kuhn, 1970. Kuhn, Thomas S. *The Structure of Scientific Revolutions, 2nd ed.* University of Chicago Press: Chicago.

Lacy and Busch, 1988. Lacy, William B., Busch, Lawrence, eds. *Biotechnology and Agricultural Cooperatives: Opportunities and Challenges*. Kentucky Agricultural Experiment Station: Lexington.

Lakatos and Musgrave, 1970. Lakatos, Imre, Musgrave, Allan eds. *Criticism and the Growth of Knowledge.* Cambridge University Press: London and New York.

Lakoff and Johnson, 1980. Lakoff, George, Johnson, Mark. *Metaphors We Live By.* The University of Chicago Press: Chicago.

Lang, 1983. Lang, John. "'Close Mystery': Wendell Berry's Poetry of Incarnation". *Renasence* 35:258-68.

Latour, 1987. Latour, Bruno. *Science in Action: How to Follow Scientists and Engineers through Society.* Harvard University Press: Cambridge.

Lears, 1981. Lears, T.J. Jackson. *No Place of Grace: Antimodernism and the Transformation of American Culture 1880-1920.* Pantheon Books: NY.

Lehning, 1973. Lehning, Arthur. "Anarchism." In *Dictionary of the History of Ideas.* Scribner's: New York.

Lepkowski, 1989a. Lepkowski, Wil. "Farmers Urged to Adopt Alternative Agriculture." *Chemical and Engineering News.* 67(11, September): 5-6.

____, 1989b. Lepkowski, Wil. "Debate Builds over Alternative Agriculture." *Chemical and Engineering News.* 67(27, November): 38-39.

Levi and Benjamin, 1977. Levi, A.M., Benjamin, A. "Focus and Flexibility in a Model of Conflict Resolution." *Journal of Conflict Resolution.* 2 (3, Summer): 405-425.

Lewis, 1955. Lewis, C.I. *The Ground and Nature of the Right.* Columbia University Press: New York.

Lewontin and Berlan, 1986. Lewontin, R.C., Berlan, Jean-Pierre. "Technology, Research and the Penetration of Capital: The Case of U.S. Agriculture." *Monthly Review.* 38: 21-34.

Lichtheim, 1967. Lichtheim, George. *The Concept of Ideology and Other Essays.* Vintage: New York.

Lipman-Blumen and Schram, 1984. Lipman-Blumen, Jean, Schram, Susan. *The Paradox of Success: The Impact on Priority Setting in Agricultural Research and Extension.* USDA: Washington, D.C.

Lipset, 1960. Lipset, Seymour Martin. "The End of Ideology? A Personal Postscript." In *Political Man: The Social Bases of Politics.* Doubleday: Garden City, NY. pp. 403-417.

Lipton and Longhurst, 1989. Lipton, Michael, Longhurst, Richard. *New Seeds and Poor People.* Johns Hopkins University Press: Baltimore.

Lockeretz, 1977. Lockeretz, William ed. *Agriculture and Energy.* Academic: New York.

____, 1983. Lockeretz, W. ed. *Environmentally Sound Agriculture: Selected Papers from the 4th Intern. Conf., INFOAM.* Praeger: New York.

Lockeretz and Wernick, 1980. Lockeretz, W., Wernick, S. "Commercial organic farming in the corn belt in comparison to conventional practices." *Rural Sociology.* 45: 709-722.

Logsdon, 1984. Logsdon, Gene. "The Importance of Traditional Farming Practices for a Sustainable Modern Agriculture." In Jackson, et al. pp. 3-18.

MacDonald, 1989. MacDonald, June Fessenden ed. *NABC Report 1, Biotechnology and Sustainable Agriculture: Policy Alternatives.* National Agricultural Biotechnology Council: Ithaca, NY.

___ 1990. MacDonald, June Fessenden ed. *NABC Report 2, Agricultural Biotechnology, Food Safety and Nutritional Quality for the Consumer.* National Agricultural Biotechnology Council: Ithaca, NY.

MacFadyen, 1984. MacFadyen, J. Tevere. *Gaining Ground: The Renewal of America's Small Farms.* Ballantine: New York.

MacIntyre, 1970. MacIntyre, Alasdair. "The Idea of a Social Science." In B. Wilson. *Rationality: Key Concepts in the Social Sciences.* Basil Blackwell: Oxford. pp. 112-130.

Madden, 1984. Madden, J. Patrick. *Regenerative Agriculture: beyond organic and sustainable food production.* CES Michigan State University: E. Lansing.

Madden and Tischbein, 1978. Madden, J.P., Tischbein, Heather. "Toward an agenda for small farm research." *American J. Agriculture Economy.* 61: 640-646.

Malinas, 1984. Malinas, Gary A. "Pesticides and Policies." *Journal of Applied Philosophy.* 1: 123-131.

Maney and Hadwiger, 1980. Maney, Ardith L., Hadwiger, Don. "Taking 'Cides: The Controversy over Agricultural Chemicals." In T. Peterson. *Farmers, Bureaucrats, and Middlemen: Historical Perspectives in American Agriculture.* Howard University Press: Washington, D.C.

Mannheim, 1951. Mannheim, Karl. *Ideology and Utopia.* Harcourt, Brace and Co.: New York.

Marcus, 1985. Marcus, Alan I. *Agricultural Science and the Quest for Legitimacy: Farmers, Agricultural Colleges, and Experiment Stations, 1870-1890.* Iowa State University Press: Ames, Iowa.

Marien, 1977. Marien, Michael. "Two visions of post-industrial society." *Futures.* 9: 415-31.

Markley and Harman, 1982. Markley, O.W., Harman, Willis W. eds. *Changing Images of Man.* From a report of the Stanford Research Institute. Pergamon: New York.

Marx, 1964. Marx, Leo. *The Machine in the Garden: Technology and the Pastoral Ideal in America.* Oxford University Press: Oxford.

Mayer and Mayer, 1973. Mayer, Andre, Mayer, Jean. "Agriculture, the island empire." *Daedalus.* (summer): pp. 83-95.

Mazur, 1981. Mazur, Alan. *The Dynamics of Technical Controversy.* Communications Press: Washington, D.C.

McKinsey, 1980. McKinsey, J. Wendell. "Contributions of the land-Grant System: Its Usefulness to the U.S. and its Potential for the Less Developed Countries." In Connor and Hessel. pp. 57-70.

Mendelsohn and Elkana, 1981. Mendelsohn, Everett, Elkana, Yehuda eds. *Sciences and Cultures: Anthropological and Historical Studies of the Sciences.* D. Reidel: Boston.

Merrill, 1976. Merrill, Richard ed. *Radical Agriculture.* Harper and Row: New York.

Meyers, 1989. Meyers, Robert. "Greening the Farm Bill." *Bioscience.* 39: 599.

Michalos, 1983. Michalos, Alex C. "Technology Assessment, Facts and Values." In Durbin, Paul T. and Rapp, Friedreich eds. *Philosophy and Technology.* D. Reidel Pub. Co.: Boston, MA. pp. 59-81.

Miller, 1984-5. Miller, Alan. "Psychosocial Origins of Conflict Over Pest Control Strategies." *Agriculture, Ecosystems, and Environment.* 12: 235-51.

Mills, 1956. Mills, C. Wright. *The Power Elite.* Oxford University Press: Oxford.

Mitchell, 1978. Mitchell, Roger L. "Agronomy in a Global Age". In Pendleton. pp. 1-6.

Mitchell, 1980. Mitchell, R.C. "How "soft", "deep", or "left"? present constituencies in the environmental movement for certain world views." *Nat. Res. J.* 20: 345-358.

Mitroff, 1974. Mitroff, Ian S. *The Subjective Side of Science: A Philosophical Inquiry into the Psychology of the Apollo Moon Scientists.* Elsevier: Amsterdam.

Moles, 1982. Moles, Jerry A. "The Future of Agriculture in Sri Lanka: Buddhism vs. Western Materialism." In Haynes and Lanier, v. 2. pp. 633-661.

Montmarquet, 1989. Montmarquet, James A. *The Idea of Agrarianism: from Hunter-Gatherer to Agrarian Radical in Western Culture.* University of Idaho Press: Moscow.

Mulkay, 1976. Mulkay, Michael. "Norms and ideology in science." *Soc. Sci. Inform.* 15: 637-656.

Mullins, 1972. Mullins, Willard A. "On the Concept of Ideology in Political Science." *American Political Science Review.* 66: 498-510.

___, 1979. Mullins, Willard A. "Truth and Ideology: Reflections on Mannheim's Paradox." *History and Theory.* 18: 141-154.

Mumford, 1966. Mumford, Lewis. "Technics and the Nature of Man." *Technology and Culture.* 7 (3): 303-17.

Murdock et al., 1990. Murdock, Steve H., Albrecht, Don E., Hamm, Rita R. "Agricultural Policy, Agricultural Sciences, and Rural Development." *J. Prod. Agric.* 3: 162-169.

Nabhan, 1984. Nabhan, Gary Paul. "Replenishing Desert Agriculture with Native Plants and their Symbionts". In Jackson, et al. pp. 172-82.

Nelkin, 1975. Nelkin, Dorothy. "The political impact of technical expertise." *Social Studies of Science.* 5: 35-54.

Newby, 1963. Newby, Idus A. "The Southern Agrarians: A View After Thirty Years." *Agricultural History.* 37.

Nissenbaum, 1980. Nissenbaum, Stephen. *Sex, Diet, and Debility in Jacksonian America: Sylvester Graham and Health Reform.* Greenwood Press: Westport, CT.

Norton, 1991. Norton, Bryan G. *Toward Unity among Environmentalists.* Oxford University Press: New York, Oxford.

Norton, et al. 1981. Norton, G.W., Fishel, W.L., Paulsen, A.A., Sundquist, W.B. eds. *Evaluation of Agricultural Research.* Proceedings of a symposium sponsored by NC-148, May 12-13, 1980. AES University of Minnesota: Minneapolis.

NRC, 1972. National Research Council, Committee on the Genetic Vulnerability of Major Crops. *Genetic Vulnerability of Major Crops.* National Academy of Science: Washington, D.C.

___, 1989. National Research Council, Committee on the Role of Alternative Farming Methods in Modern Production Agriculture. *Alternative Agriculture.* National Academy Press: Washington, D.C.

Oelhaf, 1978. Oelhaf, R.C. *Organic Agriculture: Economic and Ecological Comparisons With Conventional Methods.* Allanheld, Osmun: Montclair, NJ.

Olson et al., 1982. Olson, K.D., Langley, J., Heady, E.O. "Widespread adoption of organic farming practices: estimated impacts on U.S. agriculture." *J. of Soil and Water Cons.* 37: 41-45.

Ophuls, 1977. Ophuls, William. *Ecology and the Politics of Scarcity.* Freeman: San Francisco.

Orr, 1988. Orr, David. "Food Alchemy and Sustainable Agriculture." *Bioscience.* 38: 801-802.

Paarlberg, 1978. Paarlberg, Don. "Agriculture loves its uniqueness." *Am. J. Ag. Econ.* 60: 769-776.

___, 1980. Paarlberg, Don. *Farm and Food Policy: Issues of the 1980's.* University of Nebraska Press: Lincoln.

___, 1981. Paarlberg, Don. "The land-grant colleges and the structure issue." *Am. J. of Ag. Econ.* 63: 129-34.

Parel, 1983. Parel, Anthony ed. *Ideology, Philosophy and Politics.* Wilfred Laurier University Press: Waterloo, Ontario.

Pendleton, 1978. Pendleton, J.W. ed. *Agronomy in Today's Society.* ASA: Madison. ASA Special Pub. #33, presented at Am. Soc. of Agronomy Annual Meeting, Nov. 1977.

Perelman, 1982. Perelman, Michael. "Considerations on the Efficiency of the U.S. Food System." In Haynes and Lanier, v. 1. pp. 162-98.

Perkins, 1982. Perkins, John H. *Insects, Experts, and the Insecticide Crisis: The Quest for New Pest Management Strategies.* Plenum Pub. Co.: New York.

Peters, 1980. Peters, S. *The land in trust: a social history of the organic farming movement.* Ph.D. Thesis. Dept. of Sociology: McGill University

Peterson, 1990. Peterson, Tarla Rai. "Jefferson's Yeoman Farmer as Frontier Hero: A Self-Defeating Mythic Structure." *Agriculture and Human Values.* 7: 9-19.

Peterson et al., 1987. Peterson, E. Wesley F., Dickson, D. Bruce, Bowker, J. M. "Is the Family Farm worth Saving?" Presented to a Conference of the Society of Economic Anthropology, Riverside Ca., April 1987.

Pevear, 1982. Pevear, Richard. "On the Prose of Wendell Berry". *Hudson Review.* 35: 342-7.

Pfeiffer, 1938. Pfeiffer, E. *Biodynamic Farming and Gardening: Soil Fertility Renewal and Preservation.* Trans. Fred Heckel. Anthroposophic Press: New York.

___, 1947. Pfeiffer, E. *The Earth's Face and Human Destiny.* Rodale Press: Emmaus, PA.

Pollack, 1962. Pollack, Norman. *The Populist Response to Industrial America.* W.W. Norton: New York.

Powers and Harris, 1980. Powers, Sharon E., Harris, Craig K. "Adoption of organic farming methods: the decision-making process among Michigan farmers." Paper presented at Rural Sociology Association Annual Meetings, Ithaca, NY. August.

Putnam, 1978. Putnam, Hilary. *Meaning and the Moral Sciences.* Routledge and Kegan Paul: Boston.

Rabinow and Sullivan, 1979. Rabinow, Paul, Sullivan, William M. eds. *Interpretive Social Science: A Reader.* University of California Press: Berkeley.

Raeburn, 1990. Raeburn, Paul. "Seeds of Despair." *Issues in Science and Technology.* 6: 71-76.

Randolph and Sachs, 1981. Randolph, S. Randi, Sachs, Carolyn. "The establishment of the applied sciences: medicine and agriculture compared." In Busch. pp. 83-111.

Rasmussen, 1968. Rasmussen, Wayne D. "Advances in American agriculture: the mechanical tomato harvester as a case study." *Technology and Culture*. 9: 531-543.

Rautenstraus, 1980. Rautenstraus, R.C. "Public Responsibility of an Agronomist—a University President's View." In ASA pp. 1-6.

Raymond, 1980. Raymond, W.F. "The Relationship Between Agricultural Research and Socio-economic and Agricultural Policies." *Agriculture and Environment*. 5: 309-320.

Ricoeur, 1978. Ricoeur, Paul. "Can there be a scientific concept of ideology?" In J. Bien ed. *Phenomenology and the Social Sciences: A Dialogue*. Martinus Nijhoff: The Hague/Boston. pp. 44-59.

___, 1981. Ricoeur, Paul. *Hermeneutics and the Human Sciences*. Ed. by John B. Thompson. Cambridge University Press: Cambridge.

Rodale, 1945. Rodale, J.I. *Pay Dirt*. Rodale Press: Emmaus, PA.

Rodale, 1972. Rodale, Robert. *Sane Living in a Mad World, A Guide to the Organic Way of Life*. Rodale: Emmaus, Pa.

___, 1983. Rodale, Robert. "Breaking new ground: the search for a sustainable agriculture." *The Futurist*. 1: 15-20.

Rohde, 1963. Rohde, Gilbert. *Goals and Values underlying programs of the Farmers Union*. In *Farm Goals in Conflict*. pp. 77-86.

Rokeach, 1979. Rokeach, M. ed. *Understanding Human Values*. Free Press: New York.

Rosenberg, 1976. Rosenberg, Charles ed. *No Other Gods: On Science and American Social Thought*. Johns Hopkins University Press: Baltimore.

___, 1979. Rosenberg, Charles. "Rationalization and reality in shaping American agricultural research." In Nathan Reingold ed. *The Sciences in the American Context*. AAAS/Smithsonian Inst. Pr.: Washington, D.C.

Rossiter, 1975. Rossiter, Margaret. *The Emergence of Agricultural Science in America: Justus Liebig and the Americans, 1940-1880*. Yale University Press. New Haven, CT.

Roszak, 1972. Roszak, Theodore. *Where the Wasteland Ends: Politics and Transcendence in Postindustrial Society*. Doubleday: Garden City, NY.

Rudner, 1953. Rudner, Richard S. "The Scientist Qua Scientist Makes Value Judgements." *Philosophy of Science*. 20: 1-6.

Rushefsky, 1977. Rushefsky, Mark E. *Organic Farming: Science and Ideology in a Technological Dispute*. Ph.D. Thesis, State University of New York: Binghamton, New York.

Ruttan, 1982a. Ruttan, Vernon. *Agricultural Research Policy*. University of Minnesota Press: Minneapolis.

___, 1982b. Ruttan, Vernon. "Changing role of public and private sectors in agricultural research." *Science*. 216: 23-29.

Ruttan and Hayami, 1984. Ruttan, Vernon, Hayami, Yujiro. "Toward a Theory of Induced Institutional Innovation". *J of Development Studies*. 20: 203-223.

Ruttan et al., 1969. Ruttan, Vernon W., et al., eds. *Agricultural Policy in an Affluent Society*. W.W. Norton: New York.

Sachs, 1983. Sachs, Carolyn E. *Invisible Farmers: Women in Agricultural Production*. Allanheld, Osmun: Totowa, NJ.

Salutos and Hicks, 1951. Salutos, Theodore, Hicks, John D. *Agricultural Discontent in the Middle West, 1900-1939*. University of Wisconsin Press: Madison.

Saponara, 1988. Saponara, J. "Farm Research on Trial: Addressing the Social Costs of Agricultural Mechanization." *Science for the People*. 20: 5-9.

Schmitz and Seckler, 1970. Schmitz, Andrew, Seckler, D. "Mechanized agriculture and social welfare: the case of the tomato harvester." *Am. J. of Ag. Econ.* 52: 569-77.

Schnaiberg, 1983. Schnaiberg, Allan. "Soft Energy and Hard Labor: Structural Restraints on the Transition to Appropriate Technology." In Summers. pp. 217-234.

Schneider, 1987. Schneider, Keith. "Economics, Hate and the Farm Crisis." *New York Times*. December 7, 1987.

Schneiderman and Carpenter, 1990. Schneiderman, H. A., Carpenter, Will. "Planetary Patriotism: Sustainable Agriculture for the Future." *Environmental Science and Technology*. 24: 466-473.

Schwarz and Thompson, 1990. Schwarz, Michiel, Thompson, Michael. *Divided We Stand: Redefining Politics, Technology and Social Choice*. University of Pennsylvania Press: Philadelphia.

Scott, 1985. Scott, James C. *Weapons of the weak: everyday forms of peasant resistance*. Yale University Press: New Haven.

Shapiro, 1972. Shapiro, Edward S. "Decentralist Intellectuals and the New Deal." *Journal of American History*. 58: 938-957.

____, 1979. Shapiro, Edward S. "Catholic Agrarian Thought and the New Deal." *Catholic Historical Review*. 5

Shepard, 1985. Shepard, Philip T. "Moral Conflict in Agriculture: Conquest or Moral Coevolution?" In Edens. pp. 244-255. Also reprinted in *Agriculture and Human Values*. 1 (Fall, 1984): 17-25; and in Blatz, 1991. pp. 130-7.

____, 1988. Shepard, Philip T. "Resolving Normative Differences or Healing a 'Two-Culture' Split? A Discussion of R.D. Hollander's 'Values and Making Decisions about Agricultural Research'." *Agriculture and Human Values*. 5 (4, Fall): 79-83.

____, 1989. "Impartiality and Interpretive Intervention in Technical Controversy." In Edmund F. Byrne and Joseph C. Pitt eds. *Technological Transformation: Contextual and Conceptual Implications*. Kluwer Academic Pub.: Boston. pp. 47-65.

Shepard and Hamlin, 1987. Shepard, Philip T., Hamlin, Christopher. "How Not to Presume: Toward a Descriptive Theory of Ideology in Science and Technology Controversy." *Science, Technology and Human Values*. 12 (2, Spring): 19-28.

Shepard and Harris, 1989. Shepard, Philip T., Harris, Craig K. "Social Values, Agricultural Attitudes, and Support for Alternative Farming Systems." Presented to Second Biennial Conference of the Agriculture, Food and Human Values Society, 2-4 November, Little Rock, Arkansas.

___, 1992. Shepard, Philip T., Harris, Craig K. "What Differences Make a Difference: The Impact of Farmer's Attitudes on Choice of Farming Practices." Presented at the conference "Diversity in Food, Agriculture, Nutrition and Environment," sponsored by the Association for the Study of Food and Society and the Agriculture, Food, and Human Values Society, June 4-7, Michigan State University, East Lansing, MI.

Shi, 1985. Shi, David E. *The Simple Life: Plain Living and High Thinking in American Culture.* Oxford University Press: New York.

Shover, 1976. Shover, John L. *First Majority -- Last Minority, The Transforming of Rural Life in America.* Northern Illinois University Press: DeKalb, IL.

Smith, 1987. Smith, Tony. "Social Scientists are not Neutral Onlookers to Agricultural Policy." In Comstock. pp. 176-186.

Snyder, 1984. Snyder, Gary. "Good, Wild, Scared." In Jackson, et al. pp. 195-207.

Socolow, 1976. Socolow, Robert. "Failures of Discourse: Obstacles to the Integration of Environmental Values into Natural Resource Policy." In Laurence H. Tribe, Corinne S. Schelling, and John Voss. *When Values Conflict: Essays on Environmental Analysis, Discourse, and Decision.* Ballinger: Cambridge, MA.

Spates, 1983. Spates, James L. "The sociology of values." *Ann. Rev. Sociol.* 9: 27-49.

Speltz, 1963. Speltz, George H. *Theology of Rural Life: A Catholic Perspective.* In *Farm Goals in Conflict.* pp. 33-49.

Splinter, 1980. Splinter, W.E. "Agricultural mechanization: who wins? who loses?" *Agricultural Engineering.* 61: 14-17.

Steiner, 1924. Steiner, Rudolph. *Agriculture: A Course of Eight Lectures.* Bio-Dynamic Agric. Assoc: London.

Stewart, 1979. Stewart, Robert E., *Seven Decades that Changed America: A History of the American Society of Agricultural Engineers, 1907-1977.* America Society of Agricultural Engineers: St. Joseph, MI.

Stockdale, 1980. Stockdale, Jerry D. "Agriculture and Food Policy: Curricular Considerations". In Connor and Hessel. pp. 71-83.

Stofferahn, 1991. Stofferahn, Curtis W. et al. "Growth Fundamentalism in Dying Rural Towns: Implications for Rural Development." *Agriculture and Human Values.* 8: 25-34.

Storer, 1980. Storer, Norman William. *Science and Scientists In An Agricultural Research Organization: A Sociological Study.* Arno Press: New York.

Strange, 1982. Strange, Marty. *The Path Not Taken: A Case Study of Agricultural Research Decision-making at the Animal Science Department of the University of Neb.* Center for Rural Affairs: Walthill, Neb.

___, 1984a. Strange, Marty. "The Economic Structure of Sustainable Agriculture." In Jackson, et al. pp. 115-125.

___, 1984b. Strange, Marty ed. *It's Not All Sunshine and Fresh Air: Chronic Health Effects of Modern Farming Practices.* Center for Rural Affairs: Walthill, NB.

Summers, 1983. Summers, Gene F. *Technology and Social Change in Rural Areas: A Festschrift for Eugene A. Wilkening.* Westview: Boulder.

Suppe, 1987. Suppe, Frederick. "The Limited Applicability of Agricultural Research." *Agriculture and Human Values.* 4 (4, Fall): 4-14.

Talbot, 1983. Talbot, Allan R. *Settling Things: Six Case Studies in Environmental Mediation.* The Conservation Foundation and The Ford Foundation: Washington, D.C.

Talbot and Hadwiger, 1968. Talbot, Ross B., Hadwiger, Don F. *The Policy Process in American Agriculture.* Chandler: San Francisco.

Taylor, 1979. Taylor, Charles. *Hegel and Modern Society.* Cambridge University Press: Cambridge.

___, 1983. Taylor, Charles. "Use and Abuse of Theory." In Parel. pp. 37-59.

Thompson, 1982. Thompson, Paul B. "Risk, Ethics, and Agriculture." In Haynes and Lanier, v. 2. pp. 528-48.

Thompson, 1984. Thompson, John B. *Studies in the Theory of Ideology.* University of CA Press: Berkeley.

Thompson, 1986. Thompson, Kenneth. *Beliefs and Ideologies.* Tavistock: New York.

Thompson and Stout, 1991. Thompson, Paul B., Stout, Bill A. eds. *Beyond the Large Farm: Ethics and Research Goals for Agriculture.* Westview: Boulder.

Timmer and Nesheim, 1978. Timmer, C. Peter, Nesheim, Malden C. "Nutrition, Product Quality, and Safety." In Gardner and Richardson. pp. 155-203.

Todd, 1984. Todd, John H. "The Practice of Stewardship". In Jackson, et al. pp. 152-59.

Tribe, 1972. Tribe, Lawrence H. "Policy Science: Analysis or Ideology?" *Philosophy and Public Affairs.* 2: 66-110.

Tweeten, 1978. Tweeten, Luther. "Farm Commodity Prices and Income." In Gardner and Richardson. pp. 52-117.

___, 1983. Tweeten, Luther. "The economics of small farms." *Science.* 219: 1037-1041.

U.N. Conference on Desertification, 1977. U.N. Conference on Desertification. *Desertification: Its Causes and Consequences.* Pergamon Press: New York.

Unnevehr, 1990. Unnevehr, Laurian. "Both Urban and Rural Interests Have a Stake in the Farm Bill: A Report on Round Table Discussions of the 1990 Agricultural Legislation." *Agriculture and Human Values.* 7: 102-106.

Urwin, 1980. Urwin, Derek W. *From Ploughshare to Ballot Box: The Politics of Agrarian Defence in Europe.* Universitetsforlaget: Oslo.

U.S. Congress, 1990. U.S. Congress, Joint Economic Committee. Alternative Agriculture: Perspectives of the NAS and CAST. USGPO: Washington D.C.

U.S. Congress, House, 1980. U.S. Congress, House of Representatives. *Hearings on the Plant Variety Protection Act Amendments.* USGPO: Washington, D.C.

U.S. Congress OTA, 1981. U.S. Congress, Office of Technology Assessment. *An Assessment of the United States Food and Agricultural Research System* USGPO: Washington D.C.

___, 1985. U.S. Congress, Office of Technology Assessment. *Technology, Public Policy, and the Changing Structure of American Agriculture.* USGPO: Washington, D.C.

U.S. Congress, Senate, 1980. United States Congress Senate. *Hearings on the Plant Variety Protection Act, June 17 and 18.* USGPO: Washington D.C.

USDA, 1980. USDA, Study Team on Organic Farming. *Report and Recommendations On Organic Farming.* July. USGPO: Washington D.C.

USDA, 1981. U.S. Dept. of Agriculture. *A Time to Choose: Summary Report on the Structure of Agriculture.* Susan E. Sechler, Project Coordinator. USGPO: Washington, D.C.

U.S. General Accounting Office, 1978. United States General Accounting Office. *Changing Character and Structure of America Agriculture: An Overview.* USGAO: Washington, DC.

Usherwood, 1981. Usherwood, Noble R. ed. *Transferring Technology for Small-Scale Farming.* ASA Special Pub. #41. Am. Society for Agronomy: Madison.

U.S. National Agricultural Research and Extension Advisory Board, 1984. U.S. National Agricultural Research and Extension Advisory Board. *Science and Policy Issues: A report of citizen concerns and recommendations for agricultural research.* USGPO: Washington, D.C. July.

Vogeler, 1981.   Vogeler, Ingolf. *The Myth of the Family Farm: Agribusiness Dominance of U.S. Agriculture.* Westview: Boulder, CO.

Vogt and Albert, 1966. Vogt, Evon Z., Albert, Ethel M. eds. *People of Rimrock: A Study of Values in Five Cultures.* Harvard University Press: Cambridge, MA.

Vogtmann, 1984. Vogtmann, H. "Organic Farming Practices and Research in Europe." In Bezdicek and Power. pp. 19-36.

von Wright, 1963. von Wright, Georg Henrik. *The Varieties of Goodness.* Humanities Press: New York.

Walford, 1979.   Walford, George. *Ideologies and Their Functions: A Study in Systematic Ideology.* Villiers: London.

Walters, 1968. Walters, Charles jr. *Holding Action.* Halcyon House: Kansas City.

___, 1969. Walters, Charles jr. *Angry Testament.* Halcyon House: Kansas City.

___, 1971. Walters, Charles jr. *Unforgiven, The Biography of an Idea.* Economics Library in cooperation with Citizens Congress for Private Enterprise: Kansas City.

___, 1975. Walters, Charles jr. *The Case for Eco-Agriculture.* Acres U.S.A: Raytown, MO.

Walters and Fenzau, 1979. Walters, Charles jr, Fenzau, C. J. *An Acres U.S.A. Primer.* Acres U.S.A: Raytown, MO.

Walzer, 1967. Walzer, Michael. "On the Role of Symbolism in Political Thought." *Poli. Sci. Q.* 82: 191-204.

Ward, 1974. Ward, Colin ed. with additional material. *Peter Kropotkin: Fields, Factories and Workshops Tomorrow.* Harper & Row: New York.

Wehr, 1980a. Wehr, Elizabeth. "Diverse Critics Attack Amendment to Extend Plant Patent Protection." *Congressional Quarterly Weekly Report* 38: 1031-2.

___, 1980b. Wehr, Elizabeth, "House Agriculture Panel Approves Plant Patenting Bill." *Congressional Quarterly Weekly Report* 38: 1948.

Weinberg, 1972. Weinberg, Alvin. "Science and Trans-science." *Minerva.* 10: 209-222.

Weiss, 1977. Weiss, Carolyn. "Research for policy's sake: the enlightenment function of social research." *Policy Analysis.* 3: 531-545.

Wernick and Lockeretz, 1977. Wernick, S., Lockeretz, W. "Motivations and practical organic farmers." *Compost Science.* 18 (6): 20-24.

Whorton, 1974. Whorton, James C. *Before Silent Spring: Pesticides and Public Health in Pre-DDT America.* Princeton University Press: Princeton, NJ.

Willhelm, 1967. Willhelm, Sidney M. "A reformulation of social action theory." *Am. J. of Econ. and Sociol.* 23-31.

Willhite, 1973. Willhite, Robert G., Bowlus, Donald R., Tarbet, Donald. "An approach for resolution of attitude differences over forest management." *Environment and Behavior.* 351-366.

Williams, 1972. Williams, R. "Ideas of nature." In J. Benthall ed. *Ecology, The Shaping Enquiry.* Longman, London.

Williams, 1979. Williams, R.M. "Change and stability in values and value systems." In Rokeach. pp. 15-46.

Winner, 1977. Winner, Langdon. *Autonomous Technology.* The MIT Press: Cambridge, MA.

___, 1986. Winner, Langdon. *The Whale and the Reactor: A Search for Limits in an Age of High Technology.* The University of Chicago Press: Chicago, IL.

Wittwer, 1980. Wittwer, Sylvan. "Appropriate Agricultural Technologies and the Land-Grant Universities." Iowa State University Press: Ames. pp. 23-38.

Wojcik, 1989. Wojcik, Jan. *The Arguments of Agriculture: A Casebook in Contemporary Agricultural Controversy.* Purdue University Press: West Lafayette.

Woodward, 1959-60. Woodward, C. vann. "The Populist Heritage of the Intellectual." *American Scholar.* 29: 55-72.

Woolgar, 1988. Woolgar, Steve ed. *Knowledge and Reflexivity: New Frontiers in the Sociology of Knowledge.* Sage: London.

Worster, 1977. Worster, Donald. *Nature's Economy: The Roots of Ecology.* Anchor: Garden City, N.Y.

___, 1979. Worster, Donald. *Dust Bowl: The Southern Plains in the 1930s.* Oxford University Press: New York.

___, 1984a. Worster, Donald. "Good Farming and the Public Good." In Jackson, et al. pp. 31-41.

___, 1984b. Worster, Donald. "Thinking like a River" In Jackson, et al. pp. 56-67.

Wright, 1984. Wright, Angus. "Innocents Abroad: American Agricultural Research in Mexico". In Jackson, et al. pp. 135-51.

Young, 1978. Young, Thomas Daniel. "To Preserve so fine a Country." *Sewanee Review* 86:595-604.

Youngberg, 1978a. Youngberg, I. Garth. "Alternative agriculturalists: ideology, politics, and prospects." In Hadwiger and Browne. pp. 227-246.

___, 1978b. Youngberg, I. Garth. "The alternative agriculture movement." *Policy Studies.* 6: 524-530.

Youngberg and Buttel, 1984. Youngberg, I. Garth, Buttel, F., "Public Policy and Socio-Political Factors Affecting the Future of Sustainable Farming Systems." In Bezdicek and Power. pp. 167-85.

Index

Printed and bound by CPI Group (UK) Ltd, Croydon, CR0 4YY

23/10/2024

01778241-0007